MATHÉMATIQUES
&
APPLICATIONS

Directeurs de la collection:
J. M. Ghidaglia et P. Lascaux

24

SOCIÉTÉ DE MATHÉMATIQUES
APPLIQUÉES & INDUSTRIELLES
I.H.P. 11, rue Pierre et Marie Curie
75231 PARIS Cedex 05
Tél. (1) 44 27 66 62 - Fax (1) 44 07 03 64

Springer

Paris
Berlin
Heidelberg
New York
Barcelone
Budapest
Hong Kong
Londres
Milan
Santa Clara
Singapour
Tokyo

Philippe Destuynder Michel Salaun

Mathematical Analysis of Thin Plate Models

Springer

Philippe Destuynder
Michel Salaun
C. N. A. M. - I. A. T.
15 rue Marat
78210 St. Cyr-L'Ecole, France

Mathematics Subject Classification:
65-02 65F05 65F10 65N12 65N30 73B50 73Cxx 73C02 73C15
73C35 73K10 73K20 73M25 73V05 73V25

ISBN 978-3-540-61167-7   ISBN 978-3-642-51761-7 (eBook)
DOI 10.1007/978-3-642-51761-7

© Springer-Verlag Berlin Heidelberg 1996
Imprimé en Allemagne

SPIN: 10129987   46/3140 - 5 4 3 2 1 0 - Imprimé sur papier non acide

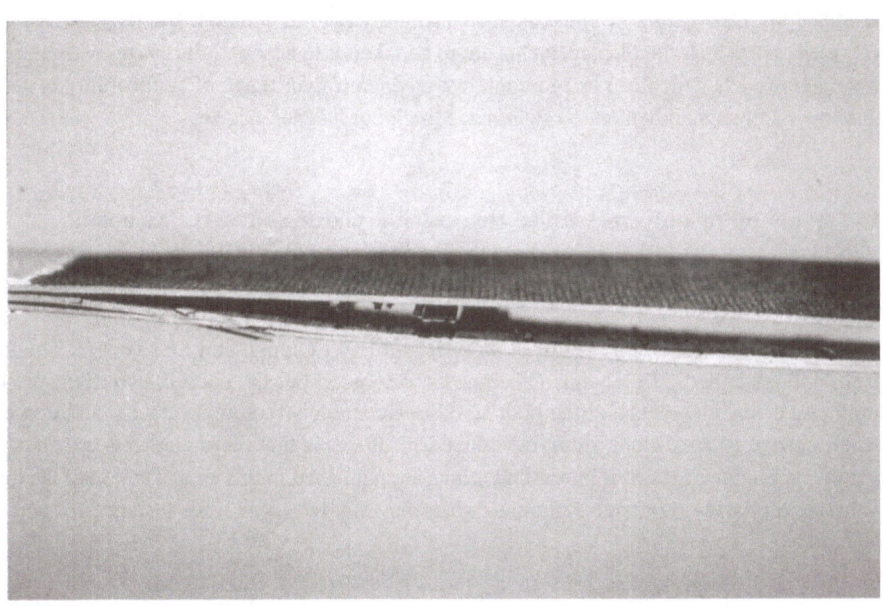

Foreword

Shells and plates have been widely studied by engineers during the last fifty years. As a matter of fact an important number of papers have been based on analytical calculations. More recently numerical simulations have been extensively used, for instance for large displacement analysis, for shape optimization or even - in linear analysis - for composite material understanding. But all these works lie on a choice of a finite element scheme which contains usually three kinds of approximations :

1. *a plate or shell model including small parameters associated to the thickness,*
2. *an approximation of the geometry (the medium surface of a shell and its boundary),*
3. *a finite element scheme in order to solve the model chosen .*

Obviously the conclusions that we can draw are very much depending on the quality of the three previous choices. For instance composite laminated plates with damage like a delamination is still an open problem even if interesting papers have already been published and based on numerical simulation using existing finite element and even plate models.

• **In our opinion the understanding of plate modelling is still an area of interest.**
Furthermore the links between the various models have to be handled with care. The certainly best understood model is the Kirchhoff-Love model which was completely justified by P.G. Ciarlet and Ph. Destuynder in linear analysis using asymptotic method. But the conclusion is not so clear as far as large displacements are to be taken into account. Furthermore there exist many other models which can be more interesting even if their range of applicability is not so well known. Let us mainly mention Reissner, Mindlin or Naghdi theories.

• **The quest for a universal finite element for plates and shells is open.**

Let us classify the existing elements into four classes :

1) – The first one was initialized by J. Argyris and R.W. Clough and J.L. Tocher. The idea was to build so-called C^1 elements. The results were good but the main drawbacks were the complexity of the elements and the poor stability for irregular solution (cracks, addition of a stiffner, change of conditions along the boundary). It seems that these elements are not used any more by engineers, even if interesting works are published in this area. They have the huge advantage to provide reference solutions when they can be applied; for instance for smooth problems.

2) - The second class is still used nowadays in the industry and is known as the non-conformal approximation. The idea is to relax the continuity of the function along the boundaries of the elements. As a matter of fact it is the continuity of the normal derivative which is relaxed. The Zienkiewicz triangle or Adini quadrangle are the most well-known elements in this class. It seems that these elements are revisited today.

3) - The third type of elements was introduced by I. Fried and then R. MacNeal and T. Hughes. They are known as the Mindlin family. But many engineers have also independently developed this kind of elements using a degenerate three dimensional approximation. The QUAD family (ie QUAD4, QUAD8 and QUAD9) are such elements. They are quadrilaterals with 4, 8 or 9 nodes. The performances are very good as far as the mesh is

not distorted. The most important drawbacks are the lack of a corresponding triangle. In practical application it is necessary to have both triangle and quadrangle elements in order to approximate a structure correctly. Additionaly some problems arise when composite materials are concerned. For instance transverse shear stresses are not correctly approximated by these elements (even if it is precisely claimed that they are relevant of a transverse shear stress theory).

4) - The last class can be understood as an extension of the previous one and very common features exist between the QUAD4 and the mixed finite elements contained in this fourth family. The leading idea is to separate the approximation of the transverse shear stress in a Mindlin plate theory from the other kinematical variables (displacements and rotations). Then two possibilities are available: continuous or discontinuous fonctions connected to the transverse shear stress. The first one is the most efficient from the mathematical point of view. The second one (ie. discontinuous fields) is the most convenient. But a difficulty appears in the vicinity of free or loaded edges. It could be also pointed out that discontinuous approximations allows a local elimination of the transverse shear variables. On the contrary, continuous fields require a mixed algorithm (as a matter of fact a penalty-duality one). But the progress obtained in programming specially with vector and parallel computers proves that the objection is meaningless, the accuracy and the stability being certainly the most important criteria to be taken into account.

This text deals only with the last class of finite elements. We had in mind to develop a finite element code which enables one to evaluate precisely what happens near free or loaded edges (junctions, ...) and to have both triangles and quadrangles facilities with very low order polynomials in order to stabilize the approximation even if singularities appear.

CONTENTS

CHAPTER I - Plate models for thin structures   page   1

   I.0 - A short description of the chapter   1

   I.1 - The three dimensionnal elastic-model   1
      I.1.1 - About the kinematics   3
      I.1.2 - About the Principle of virtual work   4
      I.1.3 - About the constitutive relationship   6
      I.1.4 - Existence uniqueness of the solution to the elastic model   7

   I.2 - The Kirchhoff-Love assumption   10

   I.3 - The Kirchhoff-Love plate model   14
      I.3.1 - Existence and uniqueness of a solution to the Kirchhoff-Love model   14
      I.3.2 - The local equations satisfied by the Kirchhoff-Love plate model   15
      I.3.3 - The transverse shear stress in Kirchhoff-Love theory   18

   I.4 - The Naghdi model revisited using mixed variational formulation   20
      I.4.1 - Existence and uniqueness of a solution to the revisited Naghdi model 21
      I.4.2 - Local equations of the Naghdi model   22

   I.5 - About the rest of the book   24

REFERENCES OF CHAPTER I   25

CHAPTER II - Variational formulations for bending plates   29

   II.0 - A brief summary of the chapter   29

   II.1 - Why a mixed formulation for plates   29

   II.2 - The primal variational formulation for Kirchhoff-Love model   29
      II.2.1 - Double Stokes formula for plates   31
      II.2.2 - The variational formulation   32
      II.2.3 - Another variational formulation   35
      II.2.4 - Interest of formulation (II.12)   40

   II.3 - The Reissner-Mindlin-Naghdi model for plates   48
      II.3.1 - The penalty method applied to the Kirchhoff-Love model   49
      II.3.2 - A correction to the penalty method   58

II.4 - Natural duality techniques for the bending plate model   73
    II.4.1 - A mixed variational formulation for Kirchhoff-Love model   75
    II.4.2 - Existence and uniqueness of solution to the mixed formulation   80
    II.4.3 - Computation of the deflection u_3 83
    II.4.4 - How to be sure we solved the right model (interpretation of the model)   84
    II.4.5 - What is the meaning of ψ and when is it zero?   85
    II.4.6 - Non-homogeneous boundary conditions   86
    II.4.7 - The revisited modified Reissner-Mindlin-Naghdi model   87
    II.4.8 - Extension to a multi-connected boundary   91

II.5 - A comparison between the mixed method and the one of section II.2.4   102

REFERENCES OF CHAPTER II   103

CHAPTER III - Finite element approximations for several plate models   105

III.0 - A summary of the chapter   105

III.1 - Basic results in finite element approximation   105
    III.1.1 - Several useful definitions   105
    III.1.2 - A brief recall concerning error estimates   110

III.2 - C^1 elements   114

III.3 - Primal finite element methods for bending plates   114

III.4 - The penalty-duality finite element method for the bending plate model   117
    III.4.1 - Stability with respect to the penalty parameter of the R.M.N. solution   118
    III.4.2 - A finite element scheme and error estimates for the R.M.N. model   141
    III.4.3 - Practical aspects in solving the R.M.N. finite element model   144
    III.4.4 - About the famous QUAD4 element   145

III.5 - Numerical approximation of the mixed formulation for a bending plate   150
    III.5.1 - General error estimates between (θ, Λ) and (θ^h, Λ^h) 153
    III.5.2 - Theoretical estimates on $u_3 - u_3{}^h$ 158
    III.5.3 - A first choice of finite elements   161
    III.5.4 - A second choice of finite elements   169

REFERENCES OF CHAPTER III   180

CHAPTER IV - Numerical tests for the mixed finite element schemes   183

IV.0 - A brief description of the chapter   183

IV.1 - Precision tests for the mixed formulation   183
    IV.1.1 - A recall of the equations to be solved

IV.1.2 - Numerical tests   185
IV.1.3 - A few remarks relative to the above numerical results   194

IV.2 Vectorial and parallel algorithms for mixed elements   194
IV.2.1 - Three strategies for solving the system (IV.18)   195
IV.2.2 - Optimization of Crout factorization   197
IV.2.3 - Optimization of node renumbering   199
IV.2.4 - Numerical tests   201

IV.3 - Concluding remarks   203

REFERENCES OF CHAPTER IV   204

CHAPTER V - A Numerical model for delamination of composite plates   207

V.0 - A brief description of the chapter   207

V.1 -What is delamination of thin multilayered plates   207

V.2 - The three-dimensional multilayered composite plate model with delamination 207

V.3 - A plate model for large delamination   210

V.4 - The three-dimensional energy release rate   216
V.4.1 - The energy release rate.   217
V.4.2 - The energy release rate for delaminated plates   219

V.5 - The mechanical example and the numerical method   224
V.5.1 - The specimen studied   224

V.6 - Concluding remarks   228

REFERENCES OF CHAPTER V   233

INDEX   235

Chapter 1

PLATE MODELS FOR THIN STRUCTURES

I.0   A short description of the chapter

In order to simplify three-dimensional equations, mechanical assumptions can be used in order to derive a simpler model for thin structures. The Kirchhoff-Love model is obtained by this method. But the Naghdi formulation is also an interesting plate model. In this chapter, both are derived from the three-dimensional theory, using the Hellinger-Reissner mixed formulation.

I.1   The three-dimensional elastic-model

Let us consider a three-dimensional body which occupies in space the open set $\Omega^\varepsilon = \omega \times]\text{-}\varepsilon, \varepsilon[$ where ω is the medium surface and ε half the thickness of the plate (see Figure I.1).

The lateral boundary of Ω^ε, which is $\Gamma^\varepsilon = \partial\omega \times]\text{-}\varepsilon, \varepsilon[$, is split into three parts. The first one is $\Gamma_0^\varepsilon = \gamma_0 \times]\text{-}\varepsilon, \varepsilon[$ and corresponds to a clamped part. The second one, denoted by $\Gamma_1^\varepsilon = \gamma_1 \times]\text{-}\varepsilon, \varepsilon[$, is such that the displacement fields are both linear through the thickness and zero in mean value (see Figure I.2). We call it a simply supported edge, but it is just a convention. The last part, $\Gamma_2^\varepsilon = \gamma_2 \times]\text{-}\varepsilon, \varepsilon[$, is a free boundary, i.e. there is no restriction at all on the displacement fields.

The forces applied to our plate can be body forces, the density of which is $f = (f_i)$ for $i = 1,2$ and 3, and surface forces $g^\pm = \left(g_i^\pm\right)$ applied on the upper and lower boundaries of the plate, say Γ_+^ε and Γ_-^ε with the definitions :

$$\Gamma_+^\varepsilon = \omega \times \{+\varepsilon\} \quad \Gamma_-^\varepsilon = \omega \times \{-\varepsilon\} \ .$$

It is also possible to consider lateral surface forces as far as they are compatible with the boundary conditions. For instance, a torque can be applied along the simply supported edge and the free boundary. Furthermore, a transverse force can also be introduced on the free edge (see Figure I.3).

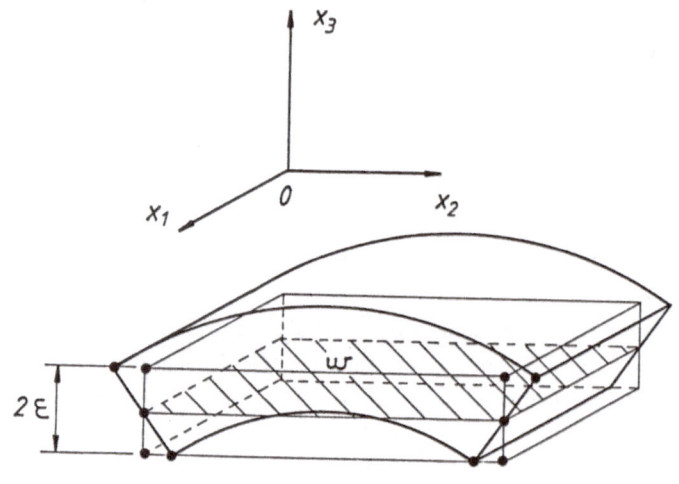

FIGURE I.1

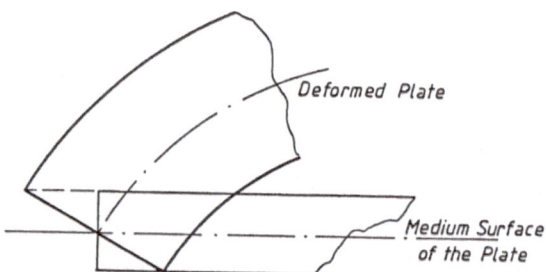

Γ_1 : SIMPLY SUPPORTED EDGE

FIGURE I.2

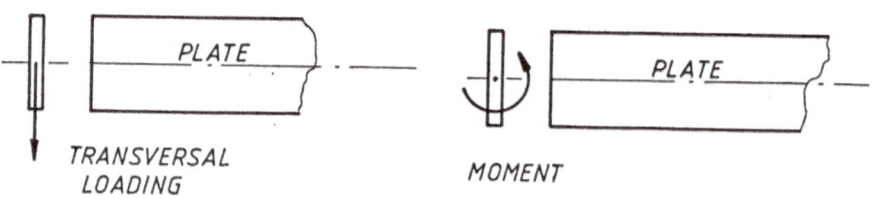

FIGURE I.3

I.1.1   About the kinematics

The plate is referred to an orthonormal set of axes $(0 ; x_1 , x_2 , x_3)$. For sake of simplicity, we assume that the medium surface ω corresponds to the plane : $x_3 = 0$. Then the displacement fields of the points of the plate are denoted by $u = (u_i)$. The linearized strains are thus :

(I.1) $$\gamma_{ij}(u) = \frac{1}{2}(\partial_i u_j + \partial_j u_i) \text{ for i and } j \in \{1, 2, 3\}$$

where ∂_i. is the partial derivative with respect to the coordinate x_i .

The admissible displacement fields for the points of the plate are such that they should be compatible with the boundary conditions and the continuity of the medium. It is convenient from now on to set (see Figure I.1 for physical explanations of the boundary conditions) :

(I.2) $V^\varepsilon = \left\{ v = (v_i) \in \left[H^1\left(\Omega^\varepsilon\right) \right]^3 \ v_i = 0 \text{ on } \Gamma^\varepsilon_0, \ v_i = x_3 \ v^1_i (x_1, x_2) \text{ with: } \sum_{\alpha = 1,2} v^1_\alpha . a_\alpha = 0 \text{ on } \gamma_1 \right\}$

a_α being the components of the unit tangent vector to γ_1, and where $H^1(\Omega^\varepsilon)$ is the Sobolev space :

$$H^1\left(\Omega^\varepsilon\right) = \left\{ f \mid f \in L^2\left(\Omega^\varepsilon\right), \partial_i f \in L^2\left(\Omega^\varepsilon\right) \right\}$$

(here, ∂_i. should be understood as a distribution derivative) equipped with the norm :

$$\left\{ \begin{array}{l} \| f \|_{1, \Omega^\varepsilon} = \left\{ |f|^2_{0, \Omega^\varepsilon} + \sum_{i = 1}^3 |\partial_i f|^2_{0, \Omega^\varepsilon} \right\}^{\frac{1}{2}}, \\[4mm] |f|_{0, \Omega^\varepsilon} = \left[\int_{\Omega^\varepsilon} |f|^2 \right]^{\frac{1}{2}}. \end{array} \right.$$

The space V^ε is a closed subspace of $\left[H^1\left(\Omega^\varepsilon\right) \right]^3$. This is almost a straightforward consequence of the continuity of the trace application from $H^1\left(\Omega^\varepsilon\right)$ into $L^2\left(\partial\omega \times \ ] \text{-}\varepsilon, \varepsilon \ [\right)$ (for details, see Appendix 1). An important result which is very much used in the mathematical theory of elasticity is the so-called Korn inequality.

Lemma I.1   Korn inequality
Let us assume that the open set ω (the medium surface of the plate) is "smooth enough"; then

the two norms on the space $\left[\mathrm{H}^1\left(\Omega^\varepsilon\right)\right]^3$:

$$\mathbf{v} = (v_i) \in \left[\mathrm{H}^1\left(\Omega^\varepsilon\right)\right]^3 \rightarrow \left\{ \sum_{i=1}^3 \|v_i\|_{1,\,\Omega^\varepsilon}^2 \right\}^{\frac{1}{2}}$$

and :

$$\mathbf{v} = (v_i) \in \left[\mathrm{H}^1\left(\Omega^\varepsilon\right)\right]^3 \rightarrow \left\{ \sum_{i=1}^3 |v_i|_{0,\,\Omega^\varepsilon}^2 + \sum_{i,\,j=1}^3 |\gamma_{ij}(v)|_{0,\,\Omega^\varepsilon}^2 \right\}^{\frac{1}{2}}$$

are equivalent.   ∎

For the proof, the reader is referred to G. Duvaut and J.L. Lions [12]. A simple consequence, but most important, is the following result.

Lemma I.2

Let us assume that ω *is "smooth enough" and that* Γ_0^ε *is such that its surface is strictly positive. Then the following semi-norm :*

$$\mathbf{v} = (v_i) \in \mathbf{V}^\varepsilon \rightarrow \left\{ \sum_{i,\,j=1}^3 |\gamma_{ij}(v)|_{0,\,\Omega^\varepsilon}^2 \right\}^{\frac{1}{2}}$$

is a norm on $\mathbf{V}^\varepsilon$ *equivalent to the one induced by the Sobolev space* $\left[\mathrm{H}^1\left(\Omega^\varepsilon\right)\right]^3$.   ∎

The proof can also be found in G. Duvaut and J.L. Lions [12].

I.1.2   About the Principle of Virtual Work

Let us denote by $\sigma = (\sigma_{ij})$ the stress tensor, which is such that : $\sigma_{ij} = \sigma_{ji}$, for all $i, j \in \{1, 2, 3\}$, then it satisfies the so-called Principle of Virtual Work :

$$\forall\, v \in \mathbf{V}^\varepsilon \,, \quad \int_{\Omega^\varepsilon} \sigma_{ij}\, \gamma_{ij}(v) = \int_{\Omega^\varepsilon} f_i\, v_i + \int_{\Gamma_+^\varepsilon \cup \Gamma_-^\varepsilon} g_i^{\pm}\, v_i \quad .$$

In the above equation, we adopt the summation convention on repeated indices. For instance :

$$\sigma_{ij}\, \gamma_{ij}(v) = \sigma_{11}\, \gamma_{11}(v) + \sigma_{22}\, \gamma_{22}(v) + \sigma_{33}\, \gamma_{33}(v) + 2\sigma_{12}\, \gamma_{12}(v) + 2\sigma_{13}\, \gamma_{13}(v) + 2\sigma_{23}\, \gamma_{23}(v) \,.$$

From now on, this convention is always used. Furthermore, the implicit summation is assumed from 1 to 2 for Greek indices, and from 1 to 3 for Latin indices. Using Stokes formula (see Appendix 1) where we choose $v \in \left[D(\Omega^\varepsilon) \right]^3$, the Principle of Virtual Work leads to the local equilibrium equations :

(I.3) $\partial_i \sigma_{ij} + f_j = 0$, for $j \in \{1, 2, 3\}$ on Ω^ε .

The boundary conditions on the lateral edge are a little bit more difficult to derive. First of all, let us assume that the stress field is smooth enough in order to justify the integration by parts. Then :

(I.4) $\forall \; v \in V^\varepsilon$, $\displaystyle\int_{\partial \Omega^\varepsilon} \sigma_{ij} \, b_i \, v_j = \int_{\Gamma^\varepsilon_+ \cup \Gamma^\varepsilon_-} g^\pm_i \, v_i$,

where $b = (b_i)$ denotes the unit outward normal to the boundary of Ω^ε . Let us split the interpretation of the above relation into three steps.

Step 1 Let $v \in V^\varepsilon$ be zero on $\Gamma^\varepsilon = \partial\omega \times \,] -\varepsilon, \varepsilon [$. Then :

$$\int_{\partial \Omega^\varepsilon} \sigma_{ij} \, b_i \, v_j = \int_{\Gamma^\varepsilon_+ \cup \Gamma^\varepsilon_-} \sigma_{ij} \, b_i \, v_j$$

and (I.4) leads to :

$$\begin{cases} \sigma_{i3} = g^+_i \; \text{on} \; \Gamma^\varepsilon_+ \, , \; i \in \{1, 2, 3\} \, , \\ \sigma_{i3} = -g^-_i \; \text{on} \; \Gamma^\varepsilon_- \, , \; i \in \{1, 2, 3\} \, . \end{cases}$$

Step 2 Let $v \in V^\varepsilon$ be zero on $\Gamma^\varepsilon_\pm$ and Γ^ε_1. As v is free on the boundary Γ^ε_2 , the relation :

$$\forall \; v \in V^\varepsilon \, , \; \int_{\Gamma^\varepsilon_2} \sigma_{ij} \, b_i \, v_j = 0$$

leads to :

(I.5) $\sigma_{ij} \, b_j = 0 \;\; \text{on} \; \Gamma^\varepsilon_2 \, , \; i \in \{1, 2, 3\} \, .$

Step 3 Let $v \in V^{\varepsilon}$ be zero on $\Gamma_{\pm}^{\varepsilon}$ and Γ_{2}^{ε}. Then one has :

$$\forall \, v \in V^{\varepsilon} \, , \, \int_{\Gamma_{1}^{\varepsilon}} \sigma_{ij} \, b_i \, v_j = 0 \quad .$$

But from the definition of V^{ε} (see (I.2)), one has on Γ_{1}^{ε}: $v_i = x_3 \, v_i^1 \, (x_1, x_2)$, where the terms v_i^1 belong to $H^1(\omega)$. Furthermore, $v_{\alpha}^1 \, a_{\alpha} = 0$ on γ_1. Thus :

$$\forall \, v_i^1 \in H^1(\omega) \text{ being zero on } \gamma_0 \cup \gamma_2 \, , \, \int_{\gamma_1} \left[\int_{-\varepsilon}^{+\varepsilon} x_3 \, \sigma_{ij} \, dx_3 \right] b_i \, v_j^1 = 0$$

which implies, because $v_{\alpha}^1 \, a_{\alpha} = 0$ on γ_1 :

$$\left[\int_{-\varepsilon}^{+\varepsilon} x_3 \, \sigma_{\alpha j} \, dx_3 \right] b_{\alpha} \, b_j = 0 \quad \text{on} \quad \gamma_1 \, , \quad \left[\int_{-\varepsilon}^{+\varepsilon} x_3 \, \sigma_{3j} \, dx_3 \right] b_j = 0 \quad \text{on} \quad \gamma_1 \, ,$$

or else, because $b_3 = 0$:

(I.6) $$\left[\int_{-\varepsilon}^{+\varepsilon} x_3 \, \sigma_{\alpha\beta} \, dx_3 \right] b_{\alpha} \, b_{\beta} = 0 \quad \text{on} \quad \gamma_1 \, , \quad \left[\int_{-\varepsilon}^{+\varepsilon} x_3 \, \sigma_{3\alpha} \, dx_3 \right] b_{\alpha} = 0 \quad \text{on} \quad \gamma_1 \, .$$

The boundary conditions (I.5) and (I.6) will be discussed in the following. From the three-dimensional point of view, they are quite natural. But for plates, the ones we obtain are different as we show later on. This is certainly one of the strangest phenomena met in plate theory. It has a lot of implications in the numerical schemes that we define in the other chapters of this book.

I.1.3   *About the constitutive relationship*

When the material, in which the plate has been built, is homogeneous and isotropic, the linear elasticity model leads to the following relationship :

(I.7) $$\sigma_{ij} = \frac{E}{1 + v} \left\{ \gamma_{ij} (u) + \frac{v}{1 - 2v} \, \gamma_{pp} (u) \, \delta_{ij} \right\}$$

where E and v are respectively Young modulus and the Poisson ratio and δ_{ij} is the Kronecker symbol (i.e. $\delta_{ij} = 0$ if $i \neq j$ and 1 otherwise). Let us recall that $E > 0$ and

$0 < v < \frac{1}{2}$. It is easy to invert relation (I.7) . This is left to the reader ! Let us just mention the result :

(I.8)
$$\gamma_{ij}(u) = \frac{1+v}{E} \sigma_{ij} - \frac{v}{E} \sigma_{pp} \delta_{ij}$$

An important property can be derived from the conditions satisfied by E and v . It is the following one :

$$\sigma_{ij}\, \gamma_{ij}(u) = \frac{E}{1+v} \left\{ \sum_{i,j=1}^{3} |\gamma_{ij}(u)|^2 + \frac{v}{1-2v}|\gamma_{pp}(u)|^2 \right\} \geq \frac{E}{1+v} \sum_{i,j=1}^{3} |\gamma_{ij}(u)|^2$$

Thus using Lemma I.2, for $v \in V^\varepsilon$, one has :

(I.9)
$$\int_{\Omega^\varepsilon} \sigma_{ij}(v) \cdot \gamma_{ij}(v) \geq c_0 ||v||^2_{V^\varepsilon} \ .$$

I.1.4   Existence, uniqueness of the solution to the elastic model

I.1.4.1. Existence and uniqueness

From the previous relations, we define the three-dimensional elasticity model as follows :

(I.10)   Find $u^\varepsilon \in V^\varepsilon$ such that $\forall \, v \in V^\varepsilon$, $a(u^\varepsilon, v) = l(v)$,

where the bilinear form $a(.,.)$ is defined by :

(I.11) $a(u,v) = \int_{\Omega^\varepsilon} \frac{E}{1+v} \left\{ \gamma_{ij}(u)\,\gamma_{ij}(v) + \frac{v}{1-2v}\,\gamma_{pp}(u)\,\gamma_{qq}(v) \right\}$

and the linear form $l(.)$ is :

(I.12) $l(v) = \int_{\Omega^\varepsilon} f_i\, v_i + \int_{\Gamma^\varepsilon_+ \cup \Gamma^\varepsilon_-} g_i^\pm\, v_i \ .$

A classical result (see G. Duvaut – J.L. Lions [12]), states that (I.10) has a unique solution. As a matter of fact, this is a direct consequence of (I.9) coupled with Lax-Milgram Theorem [12] (see Appendix 1).

1.1.4.2. Another characterization of the solution

Let us now introduce the stress space by :

$$\Sigma^\varepsilon = \left\{ \tau \mid \tau = (\tau_{ij}) \in \left[L^2(\Omega^\varepsilon) \right]^9 , \ \tau_{ij} = \tau_{ji} \right\}$$

equipped with the norm :

(I.13)
$$\| \tau \|_{\Sigma^\varepsilon} = \left\{ \sum_{i,j=1}^{3} \| \tau_{ij} \|_{0,\Omega^\varepsilon}^2 \right\}^{\frac{1}{2}} .$$

If u^ε is the solution of (I.10), we associate the stress field by :

$$\sigma_{ij}^\varepsilon = \frac{E}{1+\nu} \left\{ \gamma_{ij}(u^\varepsilon) + \frac{\nu}{1-2\nu} \gamma_{pp}(u^\varepsilon) \delta_{ij} \right\} .$$

Thus (I.10) is clearly equivalent to the Hellinger-Reissner mixed formulation [35] :

(I.14)
$$\begin{cases} \text{Find } (\sigma^\varepsilon, u^\varepsilon) \in \Sigma^\varepsilon \times V^\varepsilon \text{ such that :} \\[2mm] \forall \tau \in \Sigma^\varepsilon, \ \int_{\Omega^\varepsilon} \frac{1+\nu}{E} \sigma_{ij}^\varepsilon \tau_{ij} - \frac{\nu}{E} \sigma_{pp}^\varepsilon \tau_{qq} - \int_{\Omega^\varepsilon} \tau_{ij} \gamma_{ij}(u^\varepsilon) = 0 , \\[2mm] \forall v \in V^\varepsilon, \ \int_{\Omega^\varepsilon} \sigma_{ij}^\varepsilon \gamma_{ij}(v) = \int_{\Omega^\varepsilon} f_i v_i + \int_{\Gamma_+^\varepsilon \cup \Gamma_-^\varepsilon} g_i^\pm v_i . \end{cases}$$

The first relation is simply the variational formulation of the constitutive relationship. Furthermore, if we set for any $(\tau, v) \in \Sigma^\varepsilon \times V^\varepsilon$:

$$L(\tau, v) = \int_{\Omega^\varepsilon} \frac{1+\nu}{E} \tau_{ij} \tau_{ij} - \frac{\nu}{E} \tau_{pp} \tau_{qq} - \int_{\Omega^\varepsilon} \tau_{ij} \gamma_{ij}(v) - \int_{\Omega^\varepsilon} f_i v_i - \int_{\Gamma_+^\varepsilon \cup \Gamma_-^\varepsilon} g_i^\pm v_i ,$$

then the couple $(\sigma^\varepsilon, u^\varepsilon)$ of the space $\Sigma^\varepsilon \times V^\varepsilon$, and solution to (I.14) , is such that (see the Brezzi Theorem [4]) :

(I.15) $\forall (\tau, v) \in \Sigma^\varepsilon \times V^\varepsilon , \ L(\sigma^\varepsilon, v) \leq L(\sigma^\varepsilon, u^\varepsilon) \leq L(\tau, u^\varepsilon) .$

It could have been possible to prove directly from (I.14) the existence and uniqueness of a solution using Babuska [2] (or Brezzi, or Ladyzenskaïa [26]) Lemma, (see Appendix 1), that we recall hereafter in a simplified form, because it will be helpful in the formulation of the Kirchhoff-Love or Naghdi [29] models.

Lemma I.3

Let Σ and V be two Hilbert spaces. Let $\alpha(.,.)$ and $\beta(.,.)$ be two bilinear and continuous forms respectively on $\Sigma \times \Sigma$ and $\Sigma \times V$. We assume that there exists two strictly positive constants α_0 and β_0 such that :

$$\begin{cases} \forall \, \tau \in \Sigma, \; \alpha(\tau,\tau) \geq \alpha_0 \, \|\tau\|_\Sigma^2 \\ \forall \, v \in V, \; \sup_{\tau \in \Sigma} \frac{\beta(\tau,v)}{\|\tau\|_\Sigma} \geq \beta_0 \|v\|_V \end{cases}$$

Then for any couple $f(.)$ and $g(.)$ of linear and continuous forms on respectively Σ and V, there exists a unique solution – say (σ, u) in the space $\Sigma \times V$ - such that :

$$\begin{cases} \forall \, \tau \in \Sigma, \; \alpha(\sigma,\tau) + \beta(\tau,u) = f(\tau) \\ \forall \, v \in V, \; \beta(\sigma,v) \qquad\quad = g(v) \; . \end{cases}$$

            ■

Remark I.1

Lemma I.3 will be very useful in the whole book. But it will be used in a more complicated form for shells (which is the real Ladyzenskaïa – Babuska – Brezzi [4] theorem). In order to avoid any confusion whenever we use it, we shall recall the version we are using (see also Appendix 1).   ■

Remark I.2

The application of Lemma I.3 to (I.14) is easy if we set $\Sigma = \Sigma^\varepsilon$, $V = V^\varepsilon$ and :

$$\alpha(\sigma,\tau) = \int_{\Omega^\varepsilon} \frac{1+\nu}{E} \, \sigma_{ij} \, \tau_{ij} - \frac{\nu}{E} \, \sigma_{pp} \, \tau_{qq}$$

$$\beta(\tau,v) = -\int_{\Omega^\varepsilon} \tau_{ij} \, \gamma_{ij}(v)$$

$$f(\tau) = 0$$

$$g(v) = -l(v) = -\int_{\Omega^\varepsilon} f_i \, v_i - \int_{\Gamma_+^\varepsilon \cup \Gamma_-^\varepsilon} g_i^\pm \, v_i \; .$$

Then :

$$\forall\, \tau \in \Sigma^{\varepsilon}\,,\quad \int_{\Omega^{\varepsilon}} \frac{1+\nu}{E}\, \tau_{ij}\, \tau_{ij} - \frac{\nu}{E}\, (\tau_{pp})^2 \;=\; \int_{\Omega^{\varepsilon}} \frac{1-2\nu}{3\,E}\, \tau_{ii}\, \tau_{ij} \;+\; \int_{\Omega^{\varepsilon}} \frac{1+\nu}{E}\, \big(\tau^{D}_{ij}\, \tau^{D}_{ij}\big)$$

where $\tau^{D}_{ij} = \tau_{ij} - \frac{1}{3}\, \tau_{pp}\, \delta_{ij}$ is the deviatoric stress. Hence, noticing that $0 < 1 - 2\nu < 1 + \nu$, we obtain :

$$\forall\, \tau \in \Sigma^{\varepsilon}\,,\quad \alpha\,(\tau\,,\tau) \geq \frac{1-2\nu}{E}\, \|\tau\|^2_{\Sigma^{\varepsilon}}\,.$$

With another respect, choosing $\tau_{ij} = -\gamma_{ij}\,(v)$, we obtain :

$$\forall\, v \in V^{\varepsilon}\,,\quad \sup_{\tau\,\in\,\Sigma^{\varepsilon}} \frac{\beta\,(\tau\,,v)}{\|\tau\|_{\Sigma^{\varepsilon}}} \geq \left(\sum_{i\,,\,j\,=\,1}^{3} |\gamma_{ij}\,(v)|^2_{0\,,\,\Omega^{\varepsilon}}\right)^{\!\frac{1}{2}}\,,$$

and from Lemma I.1 :

$$\forall\, v \in V^{\varepsilon}\,,\quad \sup_{\tau\,\in\,\Sigma^{\varepsilon}} \frac{\beta\,(\tau\,,v)}{\|\tau\|_{\Sigma^{\varepsilon}}} \geq \beta_0\,\|v\|_{V^{\varepsilon}}\,.$$

The other properties (bilinearity and continuity) being quite obvious, the application of Lemma I.3 to system (I.14) is thus justified.   ∎

I.2   The Kirchhoff–Love assumption

Let us consider a very simple example of thin plate. We assume (see Figure I.1) that the structure is infinite in the direction x_1. Because the thickness is small compared to the other dimensions, Kirchhoff and Love have suggested to assimilate the plate to a collection of small pieces, each one being articulated with respect to the other and having a rigid-body behavior. It looks like these articulated wooden snakes that children have as toys. Hence the transverse shear strain remains zero, while the planar deformation is due to the articulation between the small blocks.

But this simplified description of a plate movement can be acceptable only if the components σ_{i3} of the stress field can be considered to be negligible. Following the notations of Figure I.1, these hypotheses can be formulated as follows :

(I.15) $\begin{cases} \gamma_{i3}(u) = 0 \ , \ i \in \{1,2,3\} \ \text{on} \ \omega \\ \sigma_{i3} = 0 \ , \ i \in \{1,2,3\} \ \text{on} \ \omega \ . \end{cases}$

One geometrical consequence is that **the unit normal to the medium surface remains unstretched during the deformation of the plate. Furthermore the deformed unit normal remains – during deformation – normal to the deformed medium surface**.

These "sentences" represent the Kirchhoff–Love kinematical assumption. The additional (or complementary) assumptions on the stresses ($\sigma_{i3} = 0$) are due to W.T. Koiter who was the first to clearly formulate them. As a matter of fact, even if hypotheses (I.15) are the good ones for a plate or a shell, they are not locally satisfied by the three-dimensional solution to the plate model. For instance, near a free edge or between the layers of a multilayered composite plate, the transverse shear stress plays an important role, as will be seen in Chapter V . One can say that $\sigma_{i3} = 0$ is a nice average assumption. In order to overcome these unpleasant restrictions, we shall see in Section I.3 how to define a transverse shear stress from the Kirchhoff–Love model. But the reader who is interested in a full mathematical justification of Kirchhoff–Love hypotheses should go to asymptotic methods based on the small parameters (the thickness compared to the other dimensions of the plate). The first works on this topic are due to A.L. Golden'veizer [24]. Then A. Rigolot [33], [34] used it for beams. More recently, P.G. Ciarlet and Ph. Destuynder [5], [6], gave a variational presentation of the asymptotic method, which permitted them to derive a mathematical analysis of the convergence and the error between the three-dimensional and Kirchhoff–Love solutions. For details, the reader is referred to Ph. Destuynder [16].

Extensions of this method were developed in various directions : first of all, for the Von Karman plate model by P.G. Ciarlet [7], the vibration of plates by P.G. Ciarlet and S. Kesavan [8]. The shell models are a little bit more difficult to analyze by asymptotic methods, because there are two small parameters : the new one is the ratio between the thickness and the radius of curvature. Several kinds of models can be derived depending on the relations between these two small numbers. Furthermore, a major difficulty arises from the inextensional movements which can appear on particular shells. The mathematical analysis presented in [13] is also given in a simplier form in Ph. Destuynder [16]. In the case of a non-linear shallow shell, a justification of Marguerre model has been given by P.G. Ciarlet and J.C. Paumier [9]. Let us also mention the elasto-plastic plate model (see Ph. Destuynder [14], A. Cimetière [11], and A. Cimetière and Ph. Destuynder [12] or [17]). The wave equation for plates has been studied by A. Raoult-Puech [30] and the beam model has been revisited by G. Geymonat, J.J. Marigo, F. Krasucki [23] and A. Bermudez-J. Viano [3]. But to our best knowledge, the mathematical analysis of asymptotic methods for plate models (and beams or shells) did not lead to new results (i.e. to an unknown model) except in one important case : fracture mechanics. In that case, the asymptotic analysis from the three-dimensional enabled one to derive a new expression of the energy release rate of a crack. We shall come back to this point in Chapter V with the delamination of a multi-layered plates.

Recently, P.G. Ciarlet and H. Le Dret [10], [27] have extended the asymptotic method to the analysis of junctions. This is a new area which has already allowed a clear understanding of the kinematical conditions for a junction between a three-dimensional body and a beam or a plate. Extensions to vibrations are very promising for the understanding of strange phenomena like energy concentrations and, what is very important, for the limitation of noise in complex structures (i.e. with joints).

Let us complete this section with the definition of the Kirchhoff–Love spaces respectively for stresses and displacement fields. First of all we set :

$$(I.16) \qquad \Sigma^{KL} = \left\{ \tau \mid \tau = (\tau_{ij}) \in \left[L^2(\Omega^\varepsilon) \right]^9 \ , \ \tau_{ij} = \tau_{ji} \ , \ \tau_{i3} = 0 \right\}$$

which is equipped with the norm induced by Σ^ε . Then we introduce a subspace of $\left[H^1(\Omega^\varepsilon) \right]^3$ by :

$$(I.17) \qquad V^{KL} = \left\{ v \in V^\varepsilon \mid \gamma_{i3}(v) = 0 \right\}$$

which is a subspace of V^ε equipped with the same norm. As a matter of fact, V^{KL} can be characterized through two-dimensional vector fields by the following result.

Lemma I.4

Let us introduce the space :

$$(I.18) \qquad \vartheta^{KL} = \left\{ \begin{array}{l} v = (v_i) = (\underline{v}_\alpha, v_3) \mid \underline{v}_\alpha \in H^1(\omega) \ , \ v_3 \in H^2(\omega) \\ \underline{v}_\alpha = v_3 = 0 \ \text{on} \ \gamma_0 \cup \gamma_1 \ , \ \partial_\alpha v_3 = 0 \ \text{on} \ \gamma_0 \end{array} \right\}$$

It is worth noticing that neither $\underline{v}_\alpha$ nor v_3 depend on the coordinate x_3. This space is equipped with the norm :

$$(I.19) \qquad \| v \|_{\vartheta^{KL}} = \left\{ \sum_{\alpha=1}^{2} \| \underline{v}_\alpha \|_{1,\omega}^2 + \| v_3 \|_{2,\omega}^2 \right\}^{\frac{1}{2}}$$

Then the spaces V^{KL} and ϑ^{KL} are isomorphic. ∎

Proof of Lemma I.4

Let us introduce the embedding from ϑ^{KL} into V^{KL} denoted by J and such that :

$$v = (\underline{v}_\alpha , v_3) \in \vartheta^{KL} \rightarrow J(v) = (\underline{v}_\alpha - x_3 \partial_\alpha v_3 , v_3) \in V^{KL} .$$

One has clearly $\gamma_{i3}(J(v)) = 0$ and $J(v)_i \in H^1(\Omega^\varepsilon)$ for $i \in \{ 1 , 2 , 3 \}$. Furthermore the boundary conditions on $\partial\Omega^\varepsilon$ are satisfied by $J(v)$. $J(.)$ is linear and continuous (quite obvious). In order to prove that J is an isomorphism between ϑ^{KL} and V^{KL} , it is sufficient to prove that J is onto. In other words, let v be an element of the space V^{KL} . We have to prove that v can be written as : $v = (\underline{v}_\alpha - x_3 \partial_\alpha v_3 , v_3)$. Lemma I.4 will then be a direct consequence of Banach Theorem (see for instance K. Yosida [36]).

Therefore, let us consider an element v of the space V^{KL} . We have : $\gamma_{i3}(v) = 0$ for $i \in \{ 1 , 2 , 3 \}$.

First of all, this implies : $\partial_3 v_3 = 0$, from which we deduce that v_3 can be identified with a function of the space $H^1(\omega)$.

Then the relations :

$$\gamma_{\alpha3}(v) = \frac{1}{2}(\partial_\alpha v_3 + \partial_3 v_\alpha) = 0 \quad \text{for } \alpha \in \{ 1 , 2 \} ,$$

imply that : $\partial_{33} v_\alpha = 0$, or else :

$$v_\alpha = \underline{v}_\alpha (x_1 , x_2) + x_3 b_\alpha (x_1 , x_2) .$$

where: $\underline{v}_\alpha \in H^1(\omega)$ and $b_\alpha \in H^1(\omega)$. But introducing the above expression of v_α in the equation $\gamma_{\alpha3}(v) = 0$, we obtain : $b_\alpha = - \partial_\alpha v_3$, which leads to : $v_3 \in H^2(\omega)$.

Finally :

$$\begin{cases} v_\alpha = \underline{v}_\alpha (x_1 , x_2) - x_3 \partial_\alpha v_3 \\ v_3 (x_1 , x_2) \in H^2(\omega) \end{cases}$$

and because $v \in V^{KL}$, the boundary conditions of the space ϑ^{KL} are satisfied by $\underline{v}_\alpha$ and v_3 . This completes the proof of Lemma I.4.   ∎

I.3   The Kirchhoff–Love plate model

The way we introduce this plate model is certainly unusual. But we think that it is very simple and very close to the strategy used in a finite element method. An important point is that it is not possible to use the same approach just with the displacement variational formulation given in (I.10). This is due to the fact that we need to break the constitutive relationship between strains and stresses.

The so-called Kirchhoff-Love model consists then in finding an element $\left(\sigma^{KL}, u^{KL} \right)$ of the space $\Sigma^{KL} \times V^{KL}$ such that :

$$(I.20) \qquad \begin{cases} \forall\, \tau \in \Sigma^{KL} \,,\ \alpha\left(\sigma^{KL}, \tau\right) + \beta\left(\tau, u^{KL}\right) = 0 \\ \forall\, v \in V^{KL} \,,\ \beta\left(\sigma^{KL}, v\right) = g(v), \end{cases}$$

where $\alpha\,(.,.)$, $\beta\,(.,.)$ and $g\,(.)$ are defined in section I.1.4, Remark I.2.

I.3.1   *Existence and uniqueness of a solution to the Kirchhoff-Love model*

The formulation (I.20) is rather an abstract description. We shall make it explicit in the next section. But this is not necessary for the mathematical study discussed in this one. The existence and uniqueness of a solution to (I.20) is derived from Lemma I.3. Let us check the "requirements". First of all, from :

$$\forall\, \tau \in \Sigma^{\varepsilon} \,,\ \alpha\,(\tau, \tau) \geq \alpha_0 \,\|\tau\|_{\Sigma^{\varepsilon}}^{2} \,,$$

and because $\Sigma^{KL} \subset \Sigma^{\varepsilon}$, one has immediately :

$$\forall\, \tau \in \Sigma^{KL} \,,\ \alpha\,(\tau, \tau) \geq \alpha_0 \,\|\tau\|_{\Sigma^{KL}}^{2} \,.$$

Let us now notice that for any element v in the space V^{KL}, one has :

$$\sup_{\tau \in \Sigma^{\varepsilon}} \frac{\beta\,(\tau, v)}{\|\tau\|_{\Sigma^{\varepsilon}}} = \sup_{\tau \in \Sigma^{KL}} \frac{\beta\,(\tau, v)}{\|\tau\|_{\Sigma^{KL}}}$$

because $\gamma_{i3}\,(v) = 0$. Hence we deduce that :

$$\forall\, v \in V^{KL}\,, \quad \sup_{\tau \in \Sigma^{KL}} \frac{\beta\,(\tau\,,v)}{\|\tau\|_{\Sigma^{KL}}} \geq \beta_0\; \|v\|_{V^{KL}}\;.$$

The continuity of $g(.)$ on V^{KL} being obvious as far as $g(.)$ is continuous on V^{ε}, we conclude that the Kirchhoff-Love model (I.20) has a unique solution.

I.3.2   *The local equations satisfied by the Kirchhoff-Love plate model*

There will be three steps in order to explicit formula (I.20).

Step 1 From :

$$\forall\, \tau \in \Sigma^{KL}\,, \quad \alpha\left(\sigma^{KL}, \tau\right) + \beta\left(\tau\,, u^{KL}\right) = 0\;,$$

and using definitions of $\alpha\,(.,.)$ and $\beta\,(.,.)$, we deduce that :

$$\frac{1+\nu}{E}\,\sigma^{KL}_{\alpha\beta} - \frac{\nu}{E}\,\sigma^{KL}_{\mu\mu}\,\delta_{\alpha\beta} = \gamma_{\alpha\beta}\left(u^{KL}\right) = \gamma_{\alpha\beta}\left(\underline{u}^{KL}\right) - x_3\,\partial_{\alpha\beta}\,u^{KL}_3\;,$$

where :

$$\gamma_{\alpha\beta}\left(\underline{u}^{KL}\right) = \frac{1}{2}\left(\partial_\alpha\,\underline{u}^{KL}_\beta + \partial_\beta\,\underline{u}^{KL}_\alpha\right).$$

Let us introduce the notations :

$$n^{KL}_{\alpha\beta} = \int_{-\varepsilon}^{+\varepsilon} \sigma^{KL}_{\alpha\beta}\,dx_3 \quad \text{and} \quad m^{KL}_{\alpha\beta} = \frac{3}{\varepsilon^2}\int_{-\varepsilon}^{\varepsilon} x_3\,\sigma^{KL}_{\alpha\beta}\,dx_3\;.$$

Then, assuming that E and ν are constant on Ω^{ε} (already mentioned), we deduce :

$$(\text{I.21}) \quad \begin{cases} n^{KL}_{\alpha\beta} = \dfrac{2\,E\,\varepsilon}{1-\nu^2}\left\{(1-\nu)\,\gamma_{\alpha\beta}\left(\underline{u}^{KL}\right) + \nu\,\gamma_{\mu\mu}\left(\underline{u}^{KL}\right)\delta_{\alpha\beta}\right\}, \\[2mm] m^{KL}_{\alpha\beta} = -\dfrac{2\,E\,\varepsilon}{1-\nu^2}\left\{(1-\nu)\,\partial_{\alpha\beta}\,u^{KL}_3 + \nu\,\Delta\,u^{KL}_3\,\delta_{\alpha\beta}\right\}. \end{cases}$$

Step 2 Let us set $v = (\underline{v}_\alpha\,,0) \in V^{KL}$ (see Lemma I.4). Then from the second of the two relations (I.20) - traducing the Principle of Virtual Work - we deduce :

(I.22) $\forall \, \underline{v}_\alpha \in H^1(\omega), \ \underline{v}_\alpha = 0 \text{ on } \gamma_1 \cup \gamma_0, \ \displaystyle\int_\omega n_{\alpha\beta}^{KL} \gamma_{\alpha\beta}(\underline{v}) = \int_\omega F_\alpha \underline{v}_\alpha,$

where :

(I.23) $F_\alpha = g_\alpha^+ + g_\alpha^- + \displaystyle\int_{-\varepsilon}^{+\varepsilon} f_\alpha.$

A classical variational formulation can be derived from (I.22) by eliminating $n_{\alpha\beta}^{KL}$ from the constitutive relationship (I.21) :

(I.24)
$$\left|\begin{array}{l} \text{Find } \underline{u}^{KL} \in V_t = \left\{ \underline{v} \mid \underline{v} = (\underline{v}_\alpha) \in H^1(\omega), \ \underline{v}_\alpha = 0 \text{ on } \gamma_0 \cup \gamma_1 \right\} \\ \left(V_t \text{ is equipped with the norm } H^1(\omega) \right) \text{ such that :} \\[2mm] \forall \, \underline{v} \in V_t, \ \displaystyle\int_\omega \frac{2E\varepsilon}{1-v^2} \left\{ (1-v)\gamma_{\alpha\beta}(\underline{u}^{KL})\gamma_{\alpha\beta}(\underline{v}) + v\,\gamma_{\mu\mu}(\underline{u}^{KL})\gamma_{\lambda\lambda}(\underline{v}) \right\} = \int_\omega F_\alpha \underline{v}_\alpha. \end{array}\right.$$

The existence and uniqueness of a solution to (I.24) is obtained almost exactly as for the three-dimensional case. It is based on Korn inequality (see Lemma I.2).

Using an integration by parts (called Green or Stokes formula) , one characterizes the solution of (I.24) as follows (we do not give the details) :

(I.25)
$$\left|\begin{array}{l} \partial_\beta \, n_{\alpha\beta}^{KL} + F_\alpha = 0 \text{ on } \omega \\[1mm] n_{\alpha\beta}^{KL} \, b_\beta = 0 \text{ on } \gamma_2 \\[1mm] u_\alpha^{KL} = 0 \text{ on } \gamma_1 \cup \gamma_0 \\[1mm] n_{\alpha\beta}^{KL} = \dfrac{2E\varepsilon}{1-v^2} \left\{ (1-v)\gamma_{\alpha\beta}(\underline{u}^{KL}) + v\,\gamma_{\mu\mu}(\underline{u}^{KL})\delta_{\alpha\beta} \right\} \end{array}\right.$$

which is known as the **membrane plate model**.

Step 3 Let us now choose $v = (-x_3 \, \partial_\alpha v_3, \, v_3)$ in equation (I.20), where $v_3 \in H^2(\omega)$ and satisfies the boundary conditions :

$$\left|\begin{array}{l} v_3 = 0 \text{ on } \gamma_0 \cup \gamma_1, \\[2mm] \dfrac{\partial v_3}{\partial b} = \partial_\alpha v_3 . b_\alpha = 0 \text{ on } \gamma_0. \end{array}\right.$$

Thus we obtain :

$$-\frac{\varepsilon^2}{3}\int_\omega m^{KL}_{\alpha\beta}\,\partial_{\alpha\beta}\,v_3 = \int_\omega F_3\,v_3 - \int_\omega \left\{ \varepsilon\,(g^+_\alpha - g^-_\alpha) + \int_{-\varepsilon}^{+\varepsilon} f_\alpha\,x_3\,dx_3 \right\}\partial_\alpha v_3 \quad,$$

with :

$$F_3 = g^+_3 + g^-_3 + \int_{-\varepsilon}^{+\varepsilon} f_3 \quad,$$

and :

$$m^{KL}_{\alpha\beta} = -\frac{2\,E\,\varepsilon}{1-v^2}\left\{ (1-v)\,\partial_{\alpha\beta}\,u^{KL}_3 + v\,\Delta\,u^{KL}_3\,\delta_{\alpha\beta} \right\} \quad.$$

The elimination of $m^{KL}_{\alpha\beta}$ leads to :

(I.26)

$$\forall\,v_3 \in V^0_3\ ,\quad \int_\omega \frac{2\,E\,\varepsilon^3}{3\,(1-v^2)}\left\{ (1-v)\,\partial_{\alpha\beta}\,u^{KL}_3\,\partial_{\alpha\beta}\,v_3 + v\,\Delta\,u^{KL}_3\,\Delta\,v_3 \right\}$$

$$= \int_\omega F_3\,v_3 - \int_\omega \left\{ \varepsilon\,(g^+_\alpha - g^-_\alpha) + \int_{-\varepsilon}^{+\varepsilon} f_\alpha\,x_3\,dx_3 \right\}\partial_\alpha v_3 \quad.$$

where :

$$V^0_3 = \left\{ v \in H^2(\omega)\ |\ v = 0\ \text{on}\ \gamma_0 \cup \gamma_1\ ,\ \frac{\partial v}{\partial b} = 0\ \text{on}\ \gamma_0 \right\} \quad.$$

Using a double Stokes formula, it is possible to interpret locally the solution (I.26). Let us just give the result which is justified in Chapter II :

(I.27)

$$\begin{cases} u^{KL}_3 \in V^0_3 \\[2mm] \dfrac{2\,E\,\varepsilon^3}{3\,(1-v^2)}\,\Delta^2\,u^{KL}_3 = F_3 + \partial_\alpha\left\{ \varepsilon\,(g^+_\alpha - g^-_\alpha) + \displaystyle\int_{-\varepsilon}^{+\varepsilon} f_\alpha\,x_3\,dx_3 \right\}\ \text{on}\ \omega, \\[3mm] m^{KL}_{\alpha\beta}\,b_\alpha\,b_\beta = 0\ \text{on}\ \gamma_1 \cup \gamma_2\ , \\[3mm] \dfrac{\varepsilon^2}{3}\left\{ \partial_s\left(m^{KL}_{\alpha\beta}\,a_\alpha\,b_\beta \right) + \partial_\alpha\,m_{\alpha\beta}\,b_\beta \right\} = -\left[\varepsilon\,(g^+_\alpha - g^-_\alpha) + \displaystyle\int_{-\varepsilon}^{+\varepsilon} f_\alpha\,x_3\,dx_3 \right]b_\alpha\ \text{on}\ \gamma_2, \end{cases}$$

where $\partial_s (.)$ is the derivative with respect to the curvilinear abscissa along γ_2, a_α and b_α are respectively the unit tangent vector and the unit outward normal along γ_2 (the frame (a_α, b_α) is assumed to be direct). Equations (I.27) are known as the **bending plate model**.

I.3.3   The transverse shear stress in Kirchhoff-Love theory

Let us come back to the three-dimensional stress field of the elastic model. It satisfies :

$$\forall v \in V^\varepsilon, \quad \int_{\Omega^\varepsilon} \sigma^\varepsilon_{ij} \gamma_{ij}(v) = \int_{\Omega^\varepsilon} f_i v_i + \int_{\Gamma^\varepsilon_+ \cup \Gamma^\varepsilon_-} g_i v_i ,$$

or else (if we forget the lateral boundary conditions) :

$$\begin{cases} \partial_j \sigma^\varepsilon_{ij} + f_i = 0 \text{ on } \Omega^\varepsilon, \\ \sigma^\varepsilon_{i3} = \pm g^\pm_i \text{ on } \Gamma^\varepsilon_+ \text{ and } \Gamma^\varepsilon_- . \end{cases}$$

This implies for instance :

$$\sigma^\varepsilon_{i3} = - g^-_i - \int_{-\varepsilon}^{x_3} \partial_\alpha \sigma^\varepsilon_{\alpha i} - \int_{-\varepsilon}^{x_3} f_i \quad \text{on } \omega \text{ for } i \in \{ 1, 2, 3 \} ,$$

and because σ^ε_{i3} exists, a necessary condition is :

$$0 = g^+_i + g^-_i + \int_{-\varepsilon}^{+\varepsilon} \partial_\alpha \sigma^\varepsilon_{\alpha i} + \int_{-\varepsilon}^{+\varepsilon} f_i \quad i = 1, 2, 3 .$$

Let us now assume that we know the Kirchhoff-Love approximation to the plane stress $\sigma^\varepsilon_{\alpha\beta}$ ($\alpha, \beta = 1, 2$). We obtain fromthe constitutive relationship that the plane stress is linear with respect to x_3 and therefore:

$$\sigma^{KL}_{\alpha\beta} = \frac{1}{2\varepsilon} \left(n^{KL}_{\alpha\beta} + x_3 m^{KL}_{\alpha\beta} \right) .$$

Then we deduce an approximation of the transverse shear sress by :

$$\sigma^{KL}_{\alpha 3} = - g^-_\alpha - \frac{1}{2\varepsilon} \int_{-\varepsilon}^{x_3} \partial_\beta n^{KL}_{\beta\alpha} \, dt - \frac{1}{2\varepsilon} \int_{-\varepsilon}^{x_3} t \, \partial_\beta m^{KL}_{\beta\alpha} \, dt - \int_{-\varepsilon}^{x_3} f_\alpha \, dt ,$$

or else (we used the equation satisfied by $n_{\beta\alpha}^{KL}$) :

$$(I.28) \quad \sigma_{\alpha3}^{KL} = -g_{\alpha}^{-} + \frac{(x_3 + \varepsilon)}{2\varepsilon}\left[\int_{-\varepsilon}^{+\varepsilon} f_{\alpha} \, dt + g_{\alpha}^{+} + g_{\alpha}^{-}\right] - \int_{-\varepsilon}^{x_3} f_{\alpha} \, dt - \frac{(x_3^2 - \varepsilon^2)}{4\varepsilon} \partial_{\beta} m_{\alpha\beta}^{KL} .$$

For $x = \varepsilon$, one can check that : $\sigma_{\alpha3}^{KL}(x_3 = \varepsilon) = g_{\alpha}^{+}$.

Let us now consider the component σ_{33}^{KL} . By comparison with the 3-D equilibrium relations, we set :

$$\sigma_{33}^{KL} = -g_3^- - \int_{-\varepsilon}^{x_3} \partial_{\alpha} \sigma_{\alpha3}^{KL} \, dt - \int_{-\varepsilon}^{x_3} f_3 \, dt \; ,$$

and thus, from the relation:

$$\partial_{\alpha\beta} m_{\alpha\beta}^{KL} = -\frac{3}{\varepsilon^2}\left[\int_{-\varepsilon}^{+\varepsilon} f_3 \, dt + g_3^+ + g_3^- + \varepsilon \partial_{\alpha}(g_{\alpha}^+ - g_{\alpha}^-) + \int_{-\varepsilon}^{\varepsilon} t \, \partial_{\alpha} f_{\alpha} \, dt\right] \; ,$$

we obtain :

$$\sigma_{33}^{KL} = \frac{(2\varepsilon^3 + 3\varepsilon^2 x_3 - x_3^3)}{4\varepsilon^3} \int_{-\varepsilon}^{+\varepsilon} f_3 \, dt - \int_{-\varepsilon}^{x_3} f_3 \, dt + \frac{(\varepsilon^2 - x_3^2)}{4\varepsilon} \partial_{\alpha}(g_{\alpha}^+ + g_{\alpha}^-)$$

$$+ \frac{(\varepsilon^2 x_3 - x_3^3)}{4\varepsilon^2} \partial_{\alpha}(g_{\alpha}^+ - g_{\alpha}^-) - \frac{(3\varepsilon^2 x_3 - x_3^3 + 2\varepsilon^3)}{4\varepsilon^3} \int_{-\varepsilon}^{+\varepsilon}\int_{-\varepsilon}^{t} \partial_{\alpha} f_{\alpha}$$

$$(I.29)$$

$$+ \int_{-\varepsilon}^{x_3}\int_{-\varepsilon}^{t} \partial_{\alpha} f_{\alpha} + \frac{g_3^+ - g_3^-}{2} + \frac{x_3(3\varepsilon^2 - x_3^2)}{4\varepsilon^3}(g_3^+ + g_3^-)$$

$$+ \frac{(x_3 + \varepsilon)(\varepsilon^2 - x_3^2)}{4\varepsilon^2} \int_{+\varepsilon}^{+\varepsilon} \partial_{\alpha} f_{\alpha} \; ,$$

and for $x_3 = \varepsilon$, one can check that : $\sigma_{33}^{KL}(x_3 = \varepsilon) = g_3^+$.

Both formulas (I.28) and (I.29) give an approximation of the transverse and normal stress in the plate. It can be proved that they are nice approximations to the corresponding three-dimensional stress components, inside Ω^{ε} . More precisely, it has been proved that the approximation is correct in the Sobolev space $L^2(]-\varepsilon,+\varepsilon[;H^{-1}(\omega))$ for the component $\sigma_{\alpha3}$ and $L^2(]-\varepsilon,+\varepsilon[;H^{-2}(\omega))$ for σ_{33}. But it is not valid up to the lateral boundary of the plate where a boundary layer appears (see Chapter V) . For details concerning the convergence results, see [13].

Remark I.3

One could object that our approach, concerning transverse components of the stress, is a little disappointing because we first state that $\sigma^{KL} \in \Sigma^{KL}$ and therefore $\sigma^{KL}_{\alpha3} = \sigma^{KL}_{33} = 0$. Then we construct an approximation of these terms which is different from zero. This is indeed very disturbing. But there is a justification : the transverse shear stresses are not zero, but negligible compared to the inplane components. Hence in the constitutive relationship, one can neglect the shear energy.

This property is taken into account in the variational formulation (I.20). But it can occur that locally (in the plate) the transverse shear becomes important. Then the estimates (I.28) and (I.29) can be interesting. As a matter of fact, there is a more accurate justification based on a mathematical analysis of the convergence of three-dimensional solution to a two-dimensional one. But this goes beyond the goals of this chapter. Let us refer to [13] or [16], as we already mentioned.   ∎

I.4   The Naghdi model revisited using mixed variational formulation

Let us introduce the subspaces of Σ^ε and V^ε which permit a "semi-discretization" in the direction of the thickness of the plate. First of all, we introduce the space :

$$(I.30) \quad \Sigma^{Na} = \left\{ \begin{matrix} \tau = (\tau_{ij}) \mid \tau_{\alpha\beta} = \tau^0_{\alpha\beta} + x_3\,\tau^1_{\alpha\beta} \;,\; \tau_{\alpha3} = \tau^0_{\alpha3} \;,\; \tau_{33} = 0 \\ \text{where } \tau^k_{ij} \in L^2(\omega) \text{ , so are independent on } x_3 \end{matrix} \right\}.$$

This space is equipped with the norm induced by Σ^ε, which is clearly equivalent to :

$$\|\tau\|_{\Sigma^{Na}} = \left[\sum_{k=0}^{1} \sum_{i,j=1}^{3} \|\tau^k_{ij}\|^2_{0,\omega} \right]^{\frac{1}{2}}.$$

Then we set :

$$(I.31) \quad V^{Na} = \left\{ \begin{matrix} v = (v_i) \mid v = v^0_\alpha + x_3\,v^1_\alpha \;,\; v_3 = v^0_3 \;,\; \text{where } v^k_i \in H^1(\omega) \;, \\ v^0_i = 0 \text{ on } \gamma_0 \cup \gamma_1 \;,\; v^1_\alpha\,a_\alpha = 0 \text{ on } \gamma_1 \;,\; v^1_\alpha = 0 \text{ on } \gamma_0 \end{matrix} \right\}.$$

It is quite easy to check that V^{Na} can equivalently be equipped with the norm induced by the space V^{ε} or the following one :

$$\|v\|_{V^{Na}} = \left[\sum_{k=0}^{1} \sum_{\alpha=1}^{2} \|v_\alpha^k\|_{1,\omega}^2 + \|v_3^0\|_{1,\omega}^2 \right]^{\frac{1}{2}} .$$

Let us now define the Naghdi model for plates from a variational point of view (see E. Reissner [31], [32], but also R. Mindlin [28] and P.M. Naghdi [29]) :

(I.32)
$$\begin{cases} \text{Find } \left(\sigma^{Na}, u^{Na} \right) \in \Sigma^{Na} \times V^{Na} \text{ such that :} \\ \forall\, \tau \in \Sigma^{Na} , \quad \alpha\left(\sigma^{Na}, \tau \right) + \beta\left(\tau, u^{Na} \right) = 0 , \\ \quad \forall\, v \in V^{Na} , \quad \beta\left(\sigma^{Na}, v \right) = g(v) . \end{cases}$$

($\alpha(.,.)$, $\beta(.,.)$ and $g(.)$ are defined in Remark I.2).

The reader will notice the similitude with the Kirchhoff-Love theory. The difference lies essentially in the choice of the spaces Σ^{Na} and V^{Na}. As a matter of fact, we prove in Chapter II that when the thickness tends to zero, the two models are asymptotically equivalent. An important question is : **"which one of these two models, (I.20) and (I.32), is the better for practical applications ?"** To our best knowledge, there is no correct answer. Sometimes the Kirchhoff-Love theory is better, sometimes it is the Reissner (or Mindlin or Naghdi) one. An interesting example where the latter overtakes the former is given by D. Arnold and R. Falk [1].

In our opinion, the Kirchhoff-Love theory presents few advantages and one big drawback. The advantages are mainly connected to the fact that the singular perturbation effects disappear. This is due to the small parameter, the thickness, and involves boundary layers in the three-dimensional model, which are cancelled from the Kirchhoff-Love theory. Furthermore, it is possible to construct a boundary layer theory from this model and thus it becomes, in most cases, superior to the Reissner-Mindlin-Naghdi theory. The drawback is the difficulty that one has to develop a simple finite element method for solving the model which involves a fourth-order partial derivative operator. The aim of this book is to contribute to solving this problem.

I.4.1   Existence and uniqueness of a solution to the revisited Naghdi model

Here again we use Lemma I.3. The conditions are clearly satisfied. The only one to be checked carefully is the "sup" condition :

$\exists \, \beta_0 > 0$ such that : $\forall \, v \in V^{Na}$, $\displaystyle \sup_{\tau \in \Sigma^{Na}} \frac{\beta(\tau,v)}{\|\tau\|_{\Sigma^{Na}}} \geq \beta_0 \, \|v\|_{V^{Na}}$.

In order to prove this inequality, let us notice that, from the definitions of Σ^{Na} and V^{Na} , for all $\tau \in \Sigma^{Na}$ and for all $v \in V^{Na}$, we have :

$$\beta(\tau,v) = -\left\{ 2\,\varepsilon \int_{\omega} \tau^0_{\alpha\beta} \, \partial_\alpha v^0_\beta + \frac{2\,\varepsilon^3}{3} \int_{\omega} \tau^1_{\alpha\beta} \, \partial_\alpha v^1_\beta + 2\,\varepsilon \int_{\omega} \tau^0_{\alpha 3} \left(\partial_\alpha v^0_3 + v^1_\alpha \right) \right\} \; .$$

Then, choosing for instance :

$$\tau^k_{\alpha\beta} = \tfrac{1}{2}\left(\partial_\alpha v^k_\beta + \partial_\beta v^k_\alpha \right) \; , \quad \tau^0_{\alpha 3} = \tfrac{1}{2}\left(\partial_\alpha v^0_3 + v^1_\alpha \right) \; ,$$

we deduce :

$$\forall \, v \in V^{Na} \; , \quad \sup_{\tau \in \Sigma^{Na}} \frac{\beta(\tau,v)}{\|\tau\|_{\Sigma^{Na}}} \geq C \sum_{\alpha=1}^{2} \left\{ \|v^0_\alpha\|_{1,\omega} + \|v^1_\alpha\|_{1,\omega} + \|\partial_\alpha v^0_3 + v^1_\alpha\|_{0,\omega} \right\}$$

(we used both Lemma I.2 and the ellipticity of the linear elastic operator). Finally, from :

$$\|\partial_\alpha v^0_3 + v^1_\alpha\|_{0,\omega} \geq \|\partial_\alpha v^0_3\|_{0,\omega} - \|v^1_\alpha\|_{0,\omega} \; ,$$

and because of the generalized Poincaré inequality:

$$\sum_{\alpha=1}^{2} \|\partial_\alpha v^0_3 + v^1_\alpha\|_{0,\omega} \geq c_0 \left[\|v^0_3\|_{1,\omega} - \sum_{\alpha=1,2} \|v^1_\alpha\|_{0,\omega} \right] \; ,$$

we deduce that there exists a strictly positive constant, say c_1 , such that :

$$\forall \, v \in V^{Na} \; , \quad \sup_{\tau \in \Sigma^{Na}} \frac{\beta(\tau,v)}{\|\tau\|_{\Sigma^{Na}}} \geq c_1 \|v\|_{V^{Na}} \; .$$

Hence (I.32) has a unique solution.

I.4.2 Local equations of the Naghdi model

A simple use of Stokes formula enables us to derive the following relationships from (I.32) :

$$
\text{(I.33)}\quad
\begin{cases}
\dfrac{1+\nu}{E}\,\sigma^0_{\alpha\beta} - \dfrac{\nu}{E}\,\sigma^0_{\lambda\lambda}\,\delta_{\alpha\beta} = \gamma_{\alpha\beta}\left(u^0\right) = \dfrac{1}{2}\left(\partial_\alpha u^0_\beta + \partial_\beta u^0_\alpha\right) \\[2mm]
\dfrac{1+\nu}{E}\,\sigma^1_{\alpha\beta} - \dfrac{\nu}{E}\,\sigma^1_{\lambda\lambda}\,\delta_{\alpha\beta} = \gamma_{\alpha\beta}\left(u^1\right) = \dfrac{1}{2}\left(\partial_\alpha u^1_\beta + \partial_\beta u^1_\alpha\right) \\[2mm]
\sigma^{Na}_{\alpha\beta} = \sigma^0_{\alpha\beta} + x_3\,\sigma^1_{\alpha\beta} \\[2mm]
\dfrac{2(1+\nu)}{E}\,\sigma^0_{\alpha 3} = \partial_\alpha u^0_3 + u^1_\alpha \\[2mm]
\partial_\beta \sigma^0_{\alpha\beta} + F^0_\alpha = 0 \\[2mm]
\dfrac{-\varepsilon^2}{3}\,\partial_\beta \sigma^1_{\beta\alpha} + \sigma^0_{\alpha 3} = F^1_\alpha \\[2mm]
\partial_\alpha \sigma^0_{\alpha 3} + F^0_3 = 0
\end{cases}
$$

where :

$$
\begin{cases}
F^0_\alpha = \dfrac{1}{2\varepsilon}\left(\displaystyle\int_{-\varepsilon}^{+\varepsilon} f_\alpha\,dx_3 + g^+_\alpha + g^-_\alpha\right) \\[4mm]
F^1_\alpha = \dfrac{1}{2\varepsilon}\left(\displaystyle\int_{-\varepsilon}^{+\varepsilon} x_3\,f_\alpha\,dx_3 + \varepsilon\,(g^+_\alpha - g^-_\alpha)\right) \\[4mm]
F^0_3 = \dfrac{1}{2\varepsilon}\left(\displaystyle\int_{-\varepsilon}^{+\varepsilon} f_3\,dx_3 + g^+_3 + g^-_3\right)
\end{cases}
$$

and on the boundary of ω :

$$
\begin{cases}
\sigma^1_{\alpha\beta} b_\beta b_\alpha = 0 \ \text{on}\ \gamma_1 \ , \quad \sigma^0_{\alpha\beta} b_\beta = 0 \ \text{on}\ \gamma_2 \ , \\[2mm]
\sigma^0_{\alpha 3} b_\alpha = 0 \ \text{on}\ \gamma_2 \ , \quad \sigma^1_{\alpha\beta} b_\beta = 0 \ \text{on}\ \gamma_2 \ .
\end{cases}
$$

It is worth noticing that this set of equations can be split into two separate ones as follows :

Membrane model : $\sigma^0_{\alpha\beta}$ and u^0_α are solution to :

$$
\text{(I.34)}\quad
\begin{cases}
\sigma^0_{\alpha\beta} = \dfrac{E}{1-\nu^2}\left\{(1-\nu)\,\gamma_{\alpha\beta}\left(u^0\right) + \nu\,\gamma_{\lambda\lambda}\left(u^0\right)\delta_{\alpha\beta}\right\} \ \text{on}\ \omega, \\[3mm]
\partial_\beta \sigma^0_{\alpha\beta} + F^0_\alpha = 0 \ \text{on}\ \omega, \\[2mm]
\sigma^0_{\alpha\beta} b_\beta = 0 \ \text{on}\ \gamma_2 \ , \\[2mm]
u^0_\alpha = 0 \ \text{on}\ \gamma_0 \cup \gamma_1 \ ;
\end{cases}
$$

Bending model: $\sigma^1_{\alpha\beta}$, $\sigma^0_{\alpha 3}$, u^1_α and u^0_3 are solution to:

(I.35)
$$
\begin{cases}
\sigma^1_{\alpha\beta} = \dfrac{E}{1-\nu^2}\left\{(1-\nu)\,\gamma_{\alpha\beta}(u^1) + \nu\,\gamma_{\lambda\lambda}(u^1)\,\delta_{\alpha\beta}\right\} & \text{on } \omega\,, \\[2mm]
\sigma^0_{\alpha 3} = \dfrac{E}{2(1+\nu)}\,(u^1_\alpha + \partial_\alpha u^0_3) & \text{on } \omega\,, \\[2mm]
\hspace{2cm} -\dfrac{\varepsilon^2}{3}\,\partial_\beta\,\sigma^1_{\beta\alpha} + \sigma^0_{\alpha 3} = F^1_\alpha & \text{on } \omega\,, \\[2mm]
\hspace{3cm} \partial_\alpha \sigma^0_{\alpha 3} + F^0_3 = 0 & \text{on } \omega\,,
\end{cases}
$$

with the boundary conditions:

$$
\begin{cases}
\sigma^1_{\alpha\beta} b_\beta\, b_\alpha = 0 \ \text{ on } \gamma_1\,,\ \ \sigma^1_{\alpha\beta} b_\beta = 0 \ \text{ on } \gamma_2\,, \\[2mm]
\hspace{1cm} u^0_3 = 0 \ \text{ on } \gamma_0 \cup \gamma_1\,, \\[2mm]
\hspace{1cm} u^1_\alpha = 0 \ \text{ on } \gamma_0\,,\ \ u^1_\alpha\, a_\alpha = 0 \ \text{ on } \gamma_1\,.
\end{cases}
$$

The bending model mentioned above will appear in Chapter II as a (singular!) perturbation of the Kirchhoff-Love one. This is a very important remark on which the whole book is based. We essentially believe that this singular behavior is one of the most important features in the study of bending models for plates and shells.

I.5   About the rest of the book

We derived plate models from the three-dimensional theory. A basic tool is the Hellinger-Reissner mixed formulation which enabled us to derive both the Kirchhoff-Love and Naghdi-Reissner theories. A similar strategy could be used for shells, but with a lot of additional difficulties. A brief survey of Kirchhoff-Love theory (due to W.T. Koiter) can be done . But for sake of brevity we restrict ourself to plates.

In this framework, only numerical methods for plates are presented here-after. The goal of the next chapters is to give a description, a mathematical analysis and numerical results concerning mixed finite elements for plates. The first step (Chapter II) consists in obtaining variational formulations for plates. Then the numerical analysis of finite element schemes is discussed in Chapter III and the numerical results are compared in Chapter IV. The fifth chapter is devoted to a short analysis of delamination in composite multilayered plates.

REFERENCES

[1] ARNOLD D., FALK R., [1990], The boundary layer for the Reissner - Mindlin plate model, SIAM J. Math. Anal.,26, p 1276-1290.

[2] BABUSKA B., AZIZ A.K., [1972], "Survey lectures on the mathematical foundations of the finite element method", in the Mathematical foundations of the finite element method with applications to partial differential equations. Academic Press, New York, p 3-359.

[3] BERMUDEZ A., VIANO J., [1984], Une justification des équations de la thermoélasticité des poutres à section variable par des méthodes asymptotiques RAIRO, Analyse Numérique 18, p 347-376.

[4] BREZZI F., [1974], On the existence uniqueness and approximation of saddle-point problems arising from Lagrangian multipliers ; Rev. Fr. Auto. Infor., Rech. Oper., série Rouge Anal. Numer. R-2, p 129-151.

[5] CIARLET P.G., DESTUYNDER Ph., [1979], A justification of the two-dimensional plate model, J. Mécanique 18, p 315-344.

[6] CIARLET P.G., DESTUYNDER Ph., [1979], A justification of a non linear model in plate theory, Comp Methods Appl. Mech. Engrg. 17/18, p 227-258.

[7] CIARLET P.G., [1980], A justification of the Von Karman equations, Arch. Rational Mech. Anal. 73, p 349-389.

[8] CIARLET P.G., KESAVAN S., [1981], Two dimensional approximation of three-dimensional eigenvalue problems in plate theory, Comp. Methods Appl. Mech. Engrg. 26, p 149-172.

[9] CIARLET P.G., PAUMIER J.C., [1986], A justification of the Marguerre Von-Karman equations, Comp. Mechanics 1, p 177-202.

[10] CIARLET P.G., [1990], Plates and Junctions in Elastic Multi-Structures - RMA 14, Masson, Paris.

[11] CIMETIERE A., [1984], Modèle incrémental pour la forte flexion des plaques élastoplastiques, C.R. Acad. Sci., Paris, 298, p 99-102.

[12] CIMETIERE A., DESTUYNDER Ph., [1990], A bending and buckling model for elastic-plastic plates, Eur. J. Mech. A/Solids, 9, n° 5, p 419-427.

[13] DESTUYNDER Ph., [1979], Sur une justification des modèles de plaques et de coques par les méthodes asymptotiques, Doctoral Dissertation. Univ. Paris VI.

[14] DESTUYNDER Ph., [1982], Sur les modèles de plaques minces en élastoplasticité, J. Mécanique Théorique et Appliquée 1, p 73-80.

[15] DESTUYNDER Ph., [1982], Sur la propagation des fissures dans les plaques minces en flexion, vol. 1, n° 4, p 579-594.

[16] DESTUYNDER Ph., [1986], Une théorie asymptotique des plaques minces en élasticité linéaire, RMA 2, Masson, Paris.

[17] DESTUYNDER Ph., NGUYEN Q. S., [1986], Derivation of the inelastic behavior of plates and shells from the three-dimensional models and extensions, inelastic behavior of plates and shells, Bevilacqua L., Feijoo R., Valid R., Eds., Springer, Berlin.

[18] DESTUYNDER Ph., NEVERS Th., [1987], Un modèle de calcul des forces de délaminage dans les plaques minces multicouches, vol. 6, n° 2, p 179-207.

[19] DESTUYNDER Ph., NEVERS Th., [1987], Approximation of the energy release rate in plate delamination using the asymptotic method, in Applications of Multiple Scaling in Mechanics RMA 4, Masson, Paris.

[20] DESTUYNDER Ph., [1990], Modélisation des coques minces élastiques. Physique fondamentale et appliquée, Masson, Paris.

[21] DUVAUT G., LIONS J. L., [1972], Les inéquations en mécanique et en physique, Dunod, Paris.

[22] FRIEDRICHS K. O., DRESSLER R. F., [1961], A boundary layer theory for elastic plates, C.P.A.M., 24, p 1-33.

[23] GEYMONAT G., KRASUCKI F., MARIGO J., [1987], Stress distribution in anisotropic elastic composite beams, in Applications of Multiple Scalings in Mechanics, RMA 4, Masson, Paris, p 118-133.

[24] GOLDENVEIZER A. L., [1962], Derivation of an approximate theory of bending of a

plate by the method of asymptotic integration of the equations of the theory of elasticity, J. Appl. Math. Mech. p 1000-1025.

[25] KOITER W. T., [1970], On the foundations of the linear theory for thin elastic shells, I and II. Proc. Kon. Ned Akad-Wetensch., B73, p 169-195.

[26] LADYZENSKAYA O., [1969], The mathematical theory of viscous incompressible flows, Gordon and Breach, New York.

[27] LE DRET H., [1990], Modeling of folded plate, Comput. Mech. 5, p 401-416.

[28] MINDLIN R., [1951], Influence of rototary inertia and shear on flexural motions of isotropic elastic plates, J. Appl. Mech. 18, p 31-38.

[29] NAGHDI P. M., [1972], The theory of shells and plates in Handbuch der Physik, vol. 6a 2, p 425-640 Springer, Berlin.

[30] RAOULT A., [1985], Construction d'un modèle d'évolution de plaques avec terme d'inertie de rotation, Annali di Matematica Para ed applicator 139, p 361-400.

[31] REISSNER E., [1944], On the theory of bending elastic plates, J. Math. and Phys. 23, p 184-191.

[32] REISSNER E., [1945], The effect of transverse shear deformations on the bending of elastic plates, J. Appl. Mech. 12, p 69-77.

[33] RIGOLOT A., [1972], Sur une théorie asymptotique des poutres, J. Mécanique 11, p 673-703.

[34] RIGOLOT A., [1976], Sur une théorie asymptotique des poutres droites, Doctoral Dissertation, Univ Paris VI.

[35] VALID R., [1977], La mécanique des milieux continus et le calcul des structures,. Eyrolles, Paris.

[36] YOSIDA Y., [1974], Functional Analysis, Springer, Berlin.

Chapter 2

VARIATIONAL FORMULATIONS FOR BENDING PLATES

II.0 A brief summary of the chapter

From the continuous point of view, the bending plate model involves a fourth-order operator. The goal of this chapter is to give several variational formulations of the plate model which only involve second order operators. First of all, we shall derive simple penalty methods based on the introduction of the rotation of the unit normal to the medium surface of the plate. But a major difficulty appears due to the boundary conditions on the edges of the plate. Then a more general mixed formulation is derived, for which several numerical applications are given in the other chapters. Let us go further in the details of the formulations which are explored. After a brief recall of the classical primal formulation, we focus historically on the first attempt to derive a mixed formulation. The idea is due to R. Glowinski. We extend it to more general boundary conditions.

II.1 Why a mixed formulation for plates ?

Let us start with a remark which comes from the applications. Recently multilayered composite materials have been used extensively in mechanical engineering. But a new phenomenon appeared: **delamination**. This is certainly due to transverse shear stress singularities at the interface between two layers of composites. In most cases, this damage mechanism starts from the edges of a multilayered composite plate. It is induced by an overstressing phenomenon. Then the transverse shear stress takes over. Hence a quantitative knowledge of this phenomenon is important from the mechanical engineering point of view. This is the starting point of the method we present in this chapter. But, besides this application, the interest of the variational formulations that we suggest lies in the derivation of an accurate finite element scheme which can be used for any kind of plates and shells. In order to be simple, we focus on a classical bending model in this part. The extension to composite material is given in chapter V.

II.2 The primal variational formulation for Kirchhoff-Love model

Let us consider a thin plate, the medium surface of which is denoted by ω and the thickness being 2ε. The mechanical framework that we consider is linear elasticity. The material constituting the structure is assumed to be homogeneous and isotropic (this is not a restriction as will be shown in chapter V, but just a simplification).

A transverse loading is applied, the force density of which is represented by the function f_3 (see chapter I). In addition, the lateral boundary is clamped on a part γ_0 , simply supported on another one, say γ_1 , and free on a last one, denoted γ_2 . Then, using chapter I, the Kirchhoff-Love plate model consists in finding an element u_3 which represents the deflection of the plate and such that :

(II.1)
$$\begin{cases} \dfrac{2\,E\varepsilon^3}{3\left(1 - v^2\right)}\,\Delta^2\,u_3 \ = \ f_3 \quad \text{on } \omega \\[2mm] u_3 \ = \ \dfrac{\partial\,u_3}{\partial\,b} \ = \ 0 \quad \text{on } \gamma_0 \\[2mm] u_3 \ = \ 0 \qquad m_{\alpha\beta}\,b_\alpha\,b_\beta \ = \ 0 \quad \text{on } \gamma_1 \\[2mm] m_{\alpha\beta}\,b_\alpha\,b_\beta \ = \ \partial_s\left(m_{\alpha\beta}\,a_\alpha\,b_\beta\right) + \partial_\alpha\,m_{\alpha\beta}\,b_\beta \ = \ 0 \quad \text{on } \gamma_2 \end{cases}$$

where $\{b_\alpha\}$ are the components of the unit outwards normal along the boundary of ω, and $m_{\alpha\beta}$ are the bending moments such that :

(II.2)
$$m_{\alpha\beta} \ = \ -\,\frac{2\,E\,\varepsilon}{\left(1 - v^2\right)}\,\left\{(1 - v)\,\partial_{\alpha\beta}\,u_3 \ + \ v\,\Delta u_3\,\delta_{\alpha\beta}\right\} \quad .$$

Figure II.1

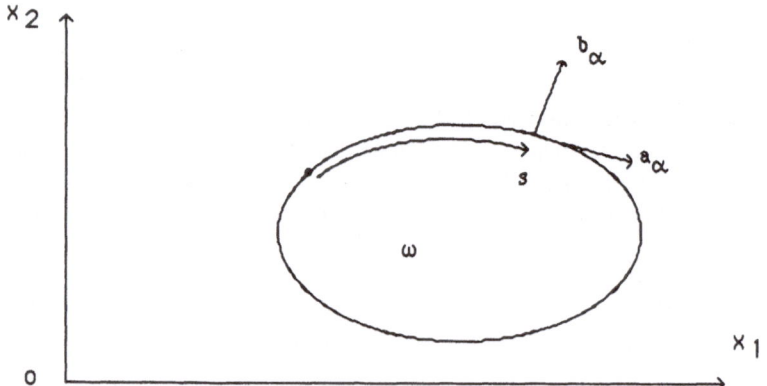

Let us recall once for all that $\dfrac{\partial u_3}{\partial b} = \partial_\alpha u_3 \, b_\alpha$ is the normal derivative and $\partial_s \, (\, . \,) = \partial_\alpha (.) \, a_\alpha = \dfrac{\partial.}{\partial s}$ is the derivative along the boundary, $\{a_\alpha\}$ denoting the components of the unit tangent to the boundary of ω. The local basis $(\{a_\alpha\}, \{b_\alpha\})$ is the one shown on Figure II.1. Hence (see figure II.1)

$$b_1 = -a_2 \quad \text{and} \quad b_2 = a_1 \; .$$

Following the notations of chapter I, the transverse shear stress through the thickness of the plate is given by the expression:

$$(\text{II}.3) \qquad \sigma_{\alpha 3} \; = \; - \frac{E\left(\epsilon^2 - x_3^2\right)}{2\left(1 - v^2\right)} \, \partial_\alpha \Delta u_3 \qquad \alpha = 1, 2$$

or else, using the bending moments:

$$(\text{II}.4) \qquad \sigma_{\alpha 3} \; = \; \frac{1}{4\,\epsilon} \left(\epsilon^2 - x_3^2\right) \partial_\beta \, m_{\alpha\beta} \quad .$$

Hence third order derivatives of u_3 are necessary in order to evaluate the transverse shear stress using formulae (II.3) or (II.4). The first variational formulation which comes straightforwardly from (II.1) and (II.2) is deduced using a double Stokes formula that we recall hereafter:

II.2.1 Double Stokes formula for plates

Let u_3 be a smooth function defined over the open set ω. The boundary of ω, say γ, is supposed to be C^1 (i.e. there exists a mapping which describes γ and is C^1). Then one has for any smooth function v_3 defined over ω, by applying Stokes formula:

$$(\text{II}.5) \qquad - \int_\omega m_{\alpha\beta} \, \partial_{\alpha\beta} \, v_3 \; = \; - \int_{\partial\omega} m_{\alpha\beta} \, b_\alpha \, \partial_\beta \, v_3 \; + \; \int_\omega \partial_\alpha \, m_{\alpha\beta} \, \partial_\beta \, v_3$$

or else applying one more time Stokes formula:

$$(\text{II}.6) \qquad - \int_\omega m_{\alpha\beta} \, \partial_{\alpha\beta} \, v_3 \; = \; - \int_{\partial\omega} m_{\alpha\beta} \, b_\alpha \, \partial_\beta \, v_3 \; + \; \int_{\partial\omega} \partial_\alpha \, m_{\alpha\beta} \, b_\beta \, v_3 \; - \; \int_\omega \partial_{\alpha\beta} \, m_{\alpha\beta} \, v_3 \; .$$

This is the so-called "double" Stokes formula for plates. There is another way to write relation

(II.6). Let us notice that on the boundary $\partial\omega$ of ω one has:

$$\partial_\beta v_3 = \frac{\partial v_3}{\partial s} a_\beta + \frac{\partial v_3}{\partial b} b_\beta$$

which is precisely the definition of the tangential (respectively normal) derivative $\dfrac{\partial v_3}{\partial s}$ (respectively $\dfrac{\partial v_3}{\partial b}$) with respect to the curvilinear abscissa s (respectively the normal coordinate along b). Hence one has from (II.6):

$$-\int_\omega m_{\alpha\beta}\, \partial_{\alpha\beta}\, v_3 = -\int_{\partial\omega} m_{\alpha\beta}\, b_\alpha\, b_\beta \frac{\partial v_3}{\partial b} - \int_{\partial\omega} m_{\alpha\beta}\, b_\alpha\, a_\beta \frac{\partial v_3}{\partial s} + \int_{\partial\omega} \partial_\alpha\, m_{\alpha\beta}\, b_\beta\, v_3$$
$$- \int_\omega \partial_{\alpha\beta}\, m_{\alpha\beta}\, v_3$$

Let us then notice that:

$$-\int_{\partial\omega} m_{\alpha\beta}\, b_\alpha\, a_\beta \frac{\partial v_3}{\partial s} = \int_{\partial\omega} \frac{\partial}{\partial s}(m_{\alpha\beta}\, b_\alpha\, a_\beta)\, v_3$$

therefore:

$$-\int_\omega m_{\alpha\beta}\, \partial_{\alpha\beta}\, v_3 = -\int_{\partial\omega} m_{\alpha\beta}\, b_\alpha\, b_\beta \frac{\partial v_3}{\partial b} + \int_{\partial\omega} \left[\frac{\partial}{\partial s}(m_{\alpha\beta}\, b_\alpha\, a_\beta) + \partial_\alpha\, m_{\alpha\beta}\, b_\beta \right] v_3$$
$$- \int_\omega \partial_{\alpha\beta}\, m_{\alpha\beta}\, v_3$$

which is the most useful expression of the "double" Stokes formula for plates.

II.2.2   The variational formulation

Let us introduce the following functional space:

(II.7) $V_3^0 = \left\{ v \mid v \in H^2(\omega); \quad v\big|_{\gamma_0 \cup \gamma_1} = 0, \quad \dfrac{\partial v}{\partial b} = 0 \quad \text{on } \gamma_0 \right\}$

It is equipped with the norm induced by the Sobolev space $H^2(\omega)$. Because the trace of a function and the trace of the derivatives are both continuous (from $H^2(\omega)$) with values respectively in the space $H^{3/2}(\partial \omega)$ and $H^{1/2}(\partial \omega)$, the space V_3^0 is a closed subspace of $H^2(\omega)$. Therefore V_3^0 is a Hilbert space.

Then assuming that u_3, the solution to (II.1) - (II.2), is smooth enough, one deduces that:

$$\forall v_3 \in V_3^0, \quad \int_\omega \frac{2 E \varepsilon^3}{3(1 - \nu^2)} \Delta^2 u_3 \, v_3 = \int_\omega f_3 \, v_3$$

But from:

$$\frac{2 E \varepsilon^3}{3(1 - \nu^2)} \Delta^2 u_3 = -\frac{\varepsilon^2}{3} \partial_{\alpha\beta} m_{\alpha\beta}$$

and using the "double" Stokes formula for plates, we deduce from (II.1) that for all $v_3 \in V_3^0$:

$$-\int_\omega m_{\alpha\beta} \, \partial_{\alpha\beta} v_3 - \int_{\partial\omega} \left(\frac{\partial}{\partial s} (m_{\alpha\beta} \, a_\alpha \, b_\beta) + \partial_\alpha \, m_{\alpha\beta} \, b_\beta \right) v_3 + \int_{\partial\omega} m_{\alpha\beta} \, b_\alpha \, b_\beta \frac{\partial v_3}{\partial b} = \frac{3}{\varepsilon^2} \int_\omega f_3 \, v_3 \ .$$

Then the boundary conditions satisfied by u_3 (and $m_{\alpha\beta}$) on $\partial\omega$ permit us to write:

(II.8) $\begin{cases} u_3 \in V_3^0 & \text{is such that:} \\ \forall v_3 \in V_3^0, \ a(u_3, v_3) = l(v_3) \end{cases}$

where:

(II.9) $\begin{cases} a(u_3, v_3) = -\dfrac{\varepsilon^2}{3} \displaystyle\int_\omega m_{\alpha\beta} \, \partial_{\alpha\beta} v_3 = \int_\omega \dfrac{2 E \varepsilon^3}{3(1 - \nu^2)} \left\{ (1 - \nu) \partial_{\alpha\beta} u_3 \, \partial_{\alpha\beta} v_3 + \nu \, \Delta u_3 \, \Delta v_3 \right\} \\[2mm] l(v_3) = \displaystyle\int_\omega f_3 \, v_3 \end{cases}$

The existence and uniqueness of a solution to system (II.8) is deduced from Lax-Milgram

Theorem [1] as soon as f_3 is regular enough. For instance $f_3 \in L^2(\omega)$ is sufficient (but not necessary). The ellipticity of the bilinear form $a(\,.\,,\,.\,)$ is obtained as follows.

From $E > 0$ and $0 < \nu < \dfrac{1}{2}$, we deduce that:

$$\forall\, v_3 \in V_3^0, \;\; a(v_3, v_3) = \int_\omega \frac{2\,E\,\varepsilon^3}{3(1-\nu^2)} \left[(1-\nu) \sum_{\alpha,\beta=1,2} \left| \partial_{\alpha\beta}\, v_3 \right|^2 + \nu \left| \Delta\, v_3 \right|^2 \right]$$

$$\geq \frac{2\,E\,\varepsilon^3}{3(1+\nu)} \sum_{\alpha,\beta=1,2} \left| \partial_{\alpha\beta}\, v_3 \right|^2_{0,\,\omega} = \frac{2\,E\,\varepsilon^3}{3(1+\nu)} \left| v_3 \right|_{2,\,\omega}$$

But the semi-norm:

$$\forall\, v_3 \in V_3^0 \;\rightarrow\; \left| v_3 \right|_{2,\,\omega}$$

is a norm on the space V_3^0, which is equivalent to the norm of $H^2(\omega)$. The proof of this last statement is very classical and can be derived in the following way. Let us assume that it is not equivalent to the one of the space $H^2(\omega)$. Then for any integer n, there exists an element v_3^n in V_3^0 such that:

$$\left| v_3^n \right|_{2,\,\omega} \leq \frac{1}{n} \left\| v_3^n \right\|_{2,\,\omega} \quad \text{and} \quad v_3^n \neq 0$$

Setting:

$$\underset{\sim}{v}_3^n = \frac{v_3^n}{\left\| v_3^n \right\|_{2,\,\omega}} \;,$$

we obtain:

$$\begin{cases} \left\| \underset{\sim}{v}_3^n \right\|_{2,\,\omega} = 1 \;, \\ \left| \underset{\sim}{v}_3^n \right|_{2,\,\omega} \leq \dfrac{1}{n} \end{cases}$$

The unit ball of a Hilbert space being (relatively) weakly compact (see for instance H. Brézis [1]), there exists a subsequence, denoted by $\underset{\sim}{v}_3^{n'}$ such that:

    i) $\underset{\sim}{v}_3^{n'} \rightarrow v_3^*$ weakly in $H^2(\omega)$,

   ii) $\underset{\sim}{v}_3^{n'} \rightarrow v_3^*$ strongly in $H^1(\omega)$,

  iii) $v_3^* \in V_3^0$,

because V_3^0 is both closed for the weak and strong topologies of the space $H^2(\omega)$. But from:

$$0 \leq \left|\underline{v}_3^{n'} - v_3^*\right|_{2,\omega}^2 = \left|\underline{v}_3^{n'}\right|_{2,\omega}^2 - 2 \sum_{\alpha,\beta=1,2} \int_\omega \partial_{\alpha\beta} \underline{v}_3^{n'} \, \partial_{\alpha\beta} v_3^* + \left|v_3^*\right|_{2,\omega}^2 ,$$

we deduce:

$$0 \leq \left|v_3^*\right|_{2,\omega}^2 \leq \lim_{n'\to\infty} \left|\underline{v}_3^{n'}\right|_{2,\omega}^2 \leq \lim_{n'\to\infty} \frac{1}{n} = 0$$

and therefore (because $|\ \ |_{2,\omega}$ is a norm on V_3^0)

$$v_3^* = 0 .$$

With another respect:

$$1 = \left\|\underline{v}_3^{n'}\right\|_{2,\omega}^2 = \left\|\underline{v}_3^{n'}\right\|_{1,\omega}^2 + \left|\underline{v}_3^{n'}\right|_{2,\omega}^2 .$$

Hence :

$$1 = \lim_{n'\to\infty} \left\|\underline{v}_3^{n'}\right\|_{1,\omega}^2 = \left\|v_3^*\right\|_{1,\omega}^2 = 0$$

which is a contradiction. So the two norms on V_3^0 (i.e. $|\ \ |_{2,\omega}$ and $\|\ \ \|_{2,\omega}$) are equivalent.   ∎

Therefore the bilinear form $a(.,.)$ introduced above is such that:

(II.10) $\exists\ C > 0,\ \forall\ v_3 \in V_3^0,\ a(v_3, v_3) \geq C \|v_3\|_{2,\omega}^2$

and the Lax-Milgram Theorem can be applied (the other hypotheses required being obvious to check). Conversely, choosing smooth enough functions v_3, one can easily check that a solution u_3 (there exists a unique one) satisfies the relations (II.1). The regularity of u_3 (i.e. $u_3 \in H^3(\omega) \cap V_3^0$ or $u_3 \in H^4(\omega) \cap V_3^0$), can be discussed from the results from P. Grisvard [2] or M. L. Williams [3].

II.2.3   Another variational formulation

Let us come back to the expression of the bilinear form a (. , .) defined at (II.9). Using Stokes formula, one obtains (for two arbitrary functions – say u and v – smooth enough):

$$a(u,v) = \int_{\partial\omega} \frac{2\,E\,\varepsilon^3}{3\,(1+v)} \partial_{\alpha\beta}\,u\,b_\beta\,\partial_\alpha\,v - \int_\omega \frac{2\,E\,\varepsilon^3}{3\,(1+v)} \partial_\alpha\,\Delta\,u\,\partial_\alpha\,v + \int_\omega \frac{2\,E\,v\,\varepsilon^3}{3\,(1-v^2)}\,\Delta\,u\,\Delta\,v$$

or else, with another use of Stokes formula:

$$a(u,v) = \int_{\partial\omega} \frac{2\,E\,\varepsilon^3}{3\,(1+v)}\left(\partial_{\alpha\beta}\,u\,b_\beta\,\partial_\alpha\,v - \Delta\,u\,b_\alpha\,\partial_\alpha\,v\right) + \int_\omega \frac{2\,E\,\varepsilon^3}{3\,(1-v^2)}\,\Delta\,u\,\Delta\,v\ .$$

From:

$$\partial_\alpha\,v = \frac{\partial\,v}{\partial\,s}\,a_\alpha + \frac{\partial\,v}{\partial\,b}\,b_\alpha \quad\text{on}\ \gamma = \partial\,\omega$$

we deduce, using local coordinates s and b on γ:

$$\partial_{\alpha\beta}\,v = \frac{\partial^2\,v}{\partial\,s^2}\,a_\alpha\,a_\beta + \frac{\partial^2\,v}{\partial\,s\,\partial\,b}\,b_\alpha\,a_\beta + \frac{\partial^2\,v}{\partial\,b\,\partial\,s}\,a_\alpha\,b_\beta$$

$$+ \frac{\partial^2\,v}{\partial\,b^2}\,b_\alpha\,b_\beta + \frac{\partial\,v}{\partial\,s}\frac{\partial\,a_\alpha}{\partial\,s}\,a_\beta + \frac{\partial\,v}{\partial\,b}\frac{\partial\,b_\alpha}{\partial\,s}\,a_\beta$$

(because the normal derivatives of a and b in the vicinity of the boundary γ are zero). But from Frenet formula (see Chapter VI):

$$\left| \begin{array}{l} \dfrac{\partial\,b_\alpha}{\partial\,s} = \dfrac{1}{R}\,a_\alpha \\[2mm] \dfrac{\partial\,a_\alpha}{\partial\,s} = -\dfrac{1}{R}\,b_\alpha \end{array} \right.$$

where R is the radius of curvature of the boundary γ of ω (counted positively along the unit outwards normal). Hence R < 0 if the domain is locally concave and R > 0 if it is locally convex. Then we obtain:

$$\partial_{\alpha\beta}\,v = \left(\frac{\partial^2\,v}{\partial\,s^2} + \frac{1}{R}\frac{\partial\,v}{\partial\,b}\right) a_\alpha\,a_\beta + \left(\frac{\partial^2\,v}{\partial\,s\,\partial\,b} - \frac{1}{R}\frac{\partial\,v}{\partial\,s}\right) b_\alpha\,a_\beta + \frac{\partial^2\,v}{\partial\,b\,\partial\,s}\,a_\alpha\,b_\beta + \frac{\partial^2\,v}{\partial\,b^2}\,b_\alpha\,b_\beta$$

Hence:

$$\begin{cases} \partial_{\alpha\beta} u \, b_{\beta} \, \partial_{\alpha} v = \dfrac{\partial^2 u}{\partial b \, \partial s} \dfrac{\partial v}{\partial s} + \dfrac{\partial^2 u}{\partial b^2} \dfrac{\partial v}{\partial b} \\[3mm] \Delta u \, b_{\alpha} \, \partial_{\alpha} v = \Delta u \dfrac{\partial v}{\partial b} = \left(\dfrac{\partial^2 u}{\partial s^2} + \dfrac{1}{R} \dfrac{\partial u}{\partial b} \right) \dfrac{\partial v}{\partial b} + \dfrac{\partial^2 u}{\partial b^2} \dfrac{\partial v}{\partial b} \end{cases}$$

which leads to:

$$a(u,v) = \int_{\omega} \frac{2 E \varepsilon^3}{3(1-v^2)} \Delta u \, \Delta v + \int_{\partial\omega} \frac{2 E \varepsilon^3}{3(1+v)} \left(\frac{\partial^2 u}{\partial b \, \partial s} \frac{\partial v}{\partial s} - \frac{\partial^2 u}{\partial s^2} \frac{\partial v}{\partial b} \right)$$

$$- \int_{\partial\omega} \frac{2 E \varepsilon^3}{3(1+v) R} \frac{\partial u}{\partial b} \frac{\partial v}{\partial b}$$

Then from a simple integration by parts on the second term in the right-hand side of the previous equality, one obtains:

$$a(u,v) = \int_{\omega} \frac{2 E \varepsilon^3}{3(1-v)^2} \Delta u \, \Delta v - \int_{\partial\omega} \frac{2 E \varepsilon^3}{3(1+v)} \left[\frac{\partial u}{\partial b} \frac{\partial^2 v}{\partial s^2} + \frac{\partial^2 u}{\partial s^2} \frac{\partial v}{\partial b} + \frac{1}{R} \frac{\partial u}{\partial b} \frac{\partial v}{\partial b} \right] \, .$$

This new expression for $a(.,.)$ has been obtained for smooth functions u and v in order to permit the integrations by parts – say for instance $C^{\infty}(\overline{\omega})$. But if we now consider two elements u and v which belong to the space $H^2(\omega)$, one has (trace Theorem):

$$v|_{\partial\omega}, \; u|_{\partial\omega} \in H^{3/2}(\partial\omega), \; \frac{\partial v}{\partial b}\Big|_{\partial\omega}, \; \frac{\partial u}{\partial b}\Big|_{\partial\omega} \in H^{1/2}(\partial\omega) \; .$$

Hence:

$$\frac{\partial^2 v}{\partial s^2}\Big|_{\partial\omega}, \; \frac{\partial^2 u}{\partial s^2}\Big|_{\partial\omega} \in H^{-1/2}(\partial\omega)$$

$(\dfrac{\partial^2 v}{\partial s^2}$ and $\dfrac{\partial^2 u}{\partial s^2}$ are distribution derivatives along the boundary $\partial\omega$).　Then $H^{-1/2}(\partial\omega)$ being the dual of $H^{1/2}(\partial\omega)$, the term

$$< \frac{\partial^2 v}{\partial s^2} , \frac{\partial u}{\partial b} > + < \frac{\partial^2 u}{\partial s^2} , \frac{\partial v}{\partial b} >$$

where $< , >$ denotes the duality between $H^{-1/2}(\partial\omega)$ and $H^{1/2}(\partial\omega)$, is perfectly defined. When u and v are smooth enough, it can be replaced by more convenient notation:

$$\int_{\partial\omega} \frac{\partial u}{\partial b}\frac{\partial^2 v}{\partial s^2} + \frac{\partial^2 u}{\partial s^2}\frac{\partial v}{\partial b}$$

Hence let us introduce the bilinear form:

$$
\begin{aligned}
(\text{II.11}) \quad \underline{a}(u,v) = &\int_{\omega} \frac{2\,E\,\varepsilon^3}{3(1-v^2)}\,\Delta u\,\Delta v - \frac{2\,E\,\varepsilon^3}{3(1+v)}\left\{ < \frac{\partial^2 v}{\partial s^2} , \frac{\partial u}{\partial b} > \right. \\
&\left. + < \frac{\partial^2 u}{\partial s^2} , \frac{\partial v}{\partial b} > \right\} - \frac{2\,E\,\varepsilon^3}{3(1+v)} \int_{\partial\omega} \frac{1}{R}\frac{\partial u}{\partial b}\frac{\partial v}{\partial b}
\end{aligned}
$$

$\underline{a}(.\,,.)$ is symmetrical, defined on $H^2(\omega) \times H^2(\omega)$ and therefore on $V_3^0 \times V_3^0$. Furthermore from (the boundary of ω is assumed to be smooth enough):

$$
\begin{aligned}
\left| < \frac{\partial^2 v}{\partial s^2} , \frac{\partial u}{\partial b} > \right| &\leq \left\| \frac{\partial^2 v}{\partial s^2} \right\|_{-1/2,\partial\omega} \left\| \frac{\partial u}{\partial b} \right\|_{1/2,\partial\omega} \\
&\leq C_1\, \|v\|_{3/2,\partial\omega}\, \|u\|_{2,\omega} \\
&\leq C_2\, \|v\|_{2,\omega}\, \|u\|_{2,\omega}
\end{aligned}
$$

(the same is true for $< \dfrac{\partial^2 u}{\partial s^2} , \dfrac{\partial v}{\partial b} >$), we deduce that $\underline{a}(.\,,.)$ is bicontinuous on $H^2(\omega) \times H^2(\omega)$. Finally, let u be an arbitrary element of the space $H^2(\omega)$. Let u^n be a sequence of $C^\infty(\overline{\omega})$ functions which tends to u in $H^2(\omega)$ as n approaches infinity. One has:

$$\underline{a}(u^n , u^n) = a(u^n , u^n) \geq C\,|u^n|^2_{2,\omega}$$

and therefore, using the continuity of $\underline{a}\,(\,.\,,\,.\,)$:

$$\underline{a}\,(u\,,\,u) \geq C|u|^2_{2\,,\,\omega}$$

Hence, V^0_3 being a subspace of $H^2\,(\omega)$:

$$\forall\ v \in V^0_3\ ,\ \underline{a}\,(v\,,\,v) \geq C\ |v|^2_{2,\omega} \geq C\ \|v\|^2_{2,\omega}$$

because of the equivalence of the two norms $|\ \ |_{2,\omega}$ and $\|\ \ \|_{2,\omega}$ on the space V^0_3 (see section II.2.2). Lax-Milgram Theorem [1] can then be applied in order to prove that the following variational equation:

(II.12) $\quad \begin{vmatrix} \text{find } u_3 \in V^0_3 \text{ such that} \\[2mm] \forall\ v_3 \in V^0_3\ ,\ \underline{a}\,(u_3\,,\,v_3)\ =\ |\,(v_3)\ \left(|\,(v_3)\ =\ \displaystyle\int_\omega f_3\ v_3 \right) \end{vmatrix}$

has a unique solution.

Remark II.1

Note that several terms can be simplified in the expression of $\underline{a}\,(\,.\,,\,.\,)$ because we use the space V^0_3. As a matter of fact $\dfrac{\partial\ u_3}{\partial\ b} = 0$ on γ_0 and $u_3 = 0$ on $\gamma_0 \cup \gamma_1$. Hence the integral

$$\int_{\partial\omega} \frac{1}{R} \frac{\partial\ u_3}{\partial\ b} \frac{\partial\ v_3}{\partial\ b}$$

is restricted to $\gamma_1 \cup \gamma_2$. ∎

Furthermore, u_3 is zero on $\gamma_0 \cup \gamma_1$. Hence $\partial^2 u_3\,/\,\partial\,s^2$ can be considered as a distribution of the space $H^{-\,1/2}\,(\gamma_2 \cup \gamma_1)$ (even if it is zero on γ_1 !). The element $\partial\,u_3/\partial\,b$ is zero on γ_0 . Hence $\partial\,u_3/\partial\,b$ is in the space $H^{1/2}_{0\,0}\,(\gamma_1 \cup \gamma_2)$. The duality between $H^{-\,1/2}\,(\partial\omega)$ and $H^{1/2}\,(\partial\omega)$ can therefore be limited to the one between $H^{-\,1/2}_{0\,0}\,(\gamma_2 \cup \gamma_1)$ and $H^{1/2}_{0\,0}\,(\gamma_1 \cup \gamma_2)$, (let us recall that the dual space of $H^{1/2}_{0\,0}\,(\gamma_1 \cup \gamma_2)$ is $H^{-\,1/2}_{0\,0}\,(\gamma_1 \cup \gamma_2)$) . One could object that $H^{-\,1/2}_{0\,0}\,(\gamma_2) \times H^{1/2}_{0\,0}\,(\gamma_2)$ would be sufficient (because $\partial^2 u_3\,/\,\partial\,s^2 = 0$ on γ_1). Unfortunately, $\partial\,u_3/\partial\,b$ does not vanish on γ_1 and there can exist concentrated distributions in the expression of $\partial^2 u_3\,/\,\partial\,s^2$, localized at the interface between γ_1 and γ_2 . This would correspond to a singularity at the interface between γ_1 and γ_2. Nevertheless, it is always possible to replace $H_{00}^{-1/2}\left(\gamma_2 \cup \gamma_1\right) \times H_{00}^{1/2}\left(\gamma_2 \cup \gamma_1\right)$ by $H_{00}^{-1/2}\left(\gamma_2^\varepsilon\right) \times H_{00}^{1/2}\left(\gamma_2^\varepsilon\right)$ where γ_2^ε is any open neighbourhood of γ_2 in $\gamma_2 \cup \gamma_1$. This is important for the practical aspects (i.e. numerical implementation).

Remark II.2

One important issue is the comparison between the two solutions to (II.8) and (II.12). Obviously they are identical, but it is worth spending some time to prove it. Let us denote by u_3 the element of V_3^0 solution to (II.12). As nothing guarantees the smoothness of u_3, we cannot use the equality:

$$a\left(u_3 , v_3\right) = \underline{a}\left(u_3 , v_3\right) \qquad \text{(even if } v_3 \text{ is smooth).}$$

Hence let us approximate u_3 by a sequence u_3^n of smooth functions belonging to the space $C^\infty\left(\overline{\omega}\right)$ with:

$$\lim_{n \to \infty} u_3^n = u_3 \quad \text{in} \quad H^2\left(\omega\right)$$

Then for any smooth $v_3 \in C^\infty\left(\overline{\omega}\right) \cap V_3^0$

$$\begin{aligned}
\underline{a}\left(u_3 , v_3\right) &= \underline{a}\left(u_3^n , v_3\right) + \underline{a}\left(u_3 - u_3^n , v_3\right) \\
&= a\left(u_3^n , v_3\right) + \underline{a}\left(u_3 - u_3^n , v_3\right) \\
&= I\left(v_3\right)
\end{aligned}$$

Then as n tends to infinity we get:

$$\forall \; v_3 \in C^\infty\left(\overline{\omega}\right) \cap V_3^0 \; , a\left(u_3 , v_3\right) = I\left(v_3\right)$$

But $C^\infty\left(\overline{\omega}\right) \cap V_3^0$ is dense in V_3^0. Finally we proved that u_3 is both solution to (II.8) and (II.12).   ∎

II.2.4   Interest of formulation (II.12)

II.2.4.1   A simply supported plate

$$\gamma_1 = \partial\omega , \gamma_0 = \gamma_2 = \varnothing \quad \text{(empty set).}$$

Then the space V_3^0 is identical to $H_0^1(\omega) \cap H^2(\omega)$. Furthermore the bilinear form $\underline{a}(\,.\,,\,.\,)$ defined at (II.11) becomes (on $V_3^0 = H_0^1(\omega) \cap H^2(\omega)$):

$$(\text{II}.13) \quad \underline{a}(u\,,\,v) = \int_\omega \frac{2\,E\,\varepsilon^3}{3\,(1-v^2)}\,\Delta\,u\,\Delta\,v - \int_{\partial\omega} \frac{2\,E\,\varepsilon^3}{3\,(1+v)\,R}\,\frac{\partial u}{\partial b}\,\frac{\partial v}{\partial b}$$

Case 1

If the boundary of ω is **piecewise linear** (i.e. polygonal) then $\frac{1}{R} = 0$ and therefore:

$$\underline{a}(u\,,\,v) = \int_\omega \frac{2\,E\,\varepsilon^3}{3\,(1-v^2)}\,\Delta\,u\,\Delta\,v\,.$$

But one should be cautious when the boundary of ω is approximated by a mesh-dependent polygonal boundary. In such a situation the boundary condition that we use here, is not correct. (see the Babuska paradox explained in P.G. Ciarlet [2]). Hence equation (II.12) is equivalent to :

$$(\text{II}.14) \quad \begin{cases} \text{find } u_3 \in H_0^1(\omega) \cap H^2(\omega) \text{ such that:} \\[2mm] \forall\ v_3 \in H_0^1(\omega) \cap H^2(\omega)\,, \displaystyle\int_\omega \frac{2\,E\,\varepsilon^3}{3\,(1-v^2)}\,\Delta\,u_3\,\Delta\,v_3 = \int_\omega f_3\,v_3 \end{cases}$$

which can be locally interpreted as follows (choosing first v_3 in $D(\omega)$ and then $v_3 \in C^\infty(\overline{\omega})$):

$$(\text{II}.15) \quad \begin{cases} \dfrac{2\,E\,\varepsilon^3}{3\,(1-v^2)}\,\Delta^2 u_3 = f_3 \text{ on } \omega\,, \\[2mm] u_3 = 0\,,\ \Delta\,u_3 = 0 \text{ on } \partial\omega\,. \end{cases}$$

These terms should be understood as distributions. Let us then set, where ψ is defined as an element of $L^2(\omega)$:

$$(\text{II}.16) \quad \begin{cases} -\,\Delta\,u_3 = \psi \text{ on } \omega\,, \\[2mm] \quad u_3 = 0 \text{ on } \partial\omega\,, \end{cases}$$

and therefore:

(II.17) $\begin{cases} -\dfrac{2\,E\,\varepsilon^3}{3\,(1-v^2)}\,\Delta\psi \;=\; f_3 \text{ on } \omega\,, \\[2mm] \psi = 0 \text{ on } \partial\omega\,. \end{cases}$

The solution to (II.14) can then be obtained using a Laplace operator with homogeneous Dirichlet boundary conditions (twice the same operator for ψ first and then u_3). Assuming the regularity of the boundary (cf. P. Grisvard [2]), we know that ψ is uniquely defined by (II.17) as an element of $H_0^1\,(\omega)$.

Case 2

The boundary of ω is an arbitrary curve, (at least piecewise C^2 in order to avoid any problem concerning the definition of the radius of curvature R). The variational equation (II.12) is equivalent to:

(II.18) $\begin{cases} \text{find } u_3 \in H_0^1\,(\omega) \cap H^2\,(\omega) \text{ such that:} \\[2mm] \forall\; v_3 \in H_0^1\,(\omega) \cap H^2\,(\omega)\,, \displaystyle\int_\omega \dfrac{2\,E\,\varepsilon^3}{3\,(1-v^2)}\,\Delta\,u_3\,\Delta\,v_3 \;-\; \displaystyle\int_{\partial\omega} \dfrac{2\,E\,\varepsilon^3}{3\,(1+v)\,R}\left(\dfrac{\partial\,u_3}{\partial b}\dfrac{\partial\,v_3}{\partial b}\right) \\[4mm] \hspace{6cm} =\; \displaystyle\int_\omega f_3\,v_3 \end{cases}$

Choosing v_3 in $D\,(\omega)$ and then in the space $C^\infty\,(\overline{\omega})$ we can locally interpret the solution to (II.18) as follows:

(II.19) $\begin{cases} \dfrac{2\,E\,\varepsilon^3}{3\,(1-v^2)}\,\Delta^2 u_3 \;=\; f_3 \text{ on } \omega\,, \\[2mm] u_3 = 0 \text{ on } \partial\omega\,, \\[2mm] \Delta\,u_3 - \dfrac{(1-v)}{R}\dfrac{\partial u_3}{\partial b} = 0 \text{ on } \partial\omega\,. \end{cases}$

Let us set:

(II.20) $\begin{cases} -\,\Delta\,u_3 = \psi \text{ on } \omega\,, \\[2mm] \hspace{0.6cm} u_3 = 0 \text{ on } \partial\omega\,, \end{cases}$

and therefore:

(II.21) $\qquad \begin{cases} -\dfrac{2\,E\,\varepsilon^3}{3\left(1-v^2\right)}\,\Delta\psi \;=\; f_3 \;\text{ on }\;\omega\,, \\[4mm] \psi \;=\; -\dfrac{(1-v)}{R}\,\dfrac{\partial u_3}{\partial b} \;\text{ on }\;\partial\omega\,. \end{cases}$

The solution to (II.12) using this decomposition is slightly tricky because the condition $\dfrac{\partial u_3}{\partial b} \in H^{1/2}(\omega)$ requires that $u_3 \in H^2(\omega)$ (as a matter of fact, this is sufficient but not necessary). Hence a regularity assumption is needed for the solution to (II.20), (the variational theory would just give $u_3 \in H_0^1(\omega)$). We refer again the reader to P. Grisvard [2] for additional details about regularity.

Let us briefly indicate how system (II.20)-(II.21) can be solved from the algorithmic point of view. First of all, let us notice that ψ satisfies the following relation:

$$\int_\omega \psi \;=\; -\int_\omega \Delta\,u_3 \;=\; -\int_{\partial\omega} \frac{\partial\,u_3}{\partial b} \;=\; \frac{1}{1-v}\int_{\partial\omega} R\,\psi$$

or else

(II.22) $\qquad \displaystyle\int_\omega \psi \;-\; \frac{1}{1-v}\int_{\partial\omega} R\,\psi \;=\; 0\,.$

Then we assume that for any function h of the space $L^2(\omega)$ the unique solution to the Poisson equation (this is always true because of the boundary regularity that we have assumed):

$$\begin{cases} -\,\Delta z \;=\; h \quad\text{on}\quad \omega \\[2mm] z \;=\; 0 \quad\text{on}\quad \partial\omega \\[2mm] z \in H^1(\omega) \end{cases}$$

belongs to the space $H^2(\omega)$, and that there exists a constant c such that:

(II.23) $\qquad \|z\|_{2,\omega} \;\leq\; C\,\|h\|_{0,\omega}\;\;;$

(Obviously C is supposed to be independent of the function h). Let us now consider an element g of the space $H^{-1/2}(\partial\omega)$ such that:

$$\frac{2\,E\,\varepsilon^3}{3(1-v^2)}\langle g\,,\,1\rangle \;+\; \int_\omega f_3 \;=\; 0$$

$(<, >)$ is here the duality between $H^{-1/2}(\partial\omega)$ and $H^{1/2}(\partial\omega)$. We associate the element ψ of the space $H^1(\omega)$ such that:

(II.24)
$$\begin{cases} -\dfrac{2\,E\,\varepsilon^3}{3\left(1-v^2\right)}\,\Delta\psi = f_3 \text{ on } \omega, \\[3mm] \dfrac{\partial\psi}{\partial b} = g \text{ on } \partial\omega. \end{cases}$$

We deduce from Fredholm alternative that ψ is perfectly defined up to a constant by system (II.24). This constant is fixed by condition (II.22) :

(II.25)
$$\int_\omega \psi - \frac{1}{1-v}\int_{\partial\omega} R\,\psi = 0 \quad .$$

This is always possible if and only if:

$$\int_\omega 1 - \frac{1}{1-v}\int_{\partial\omega} R = \text{mes}\,(\omega) - \frac{1}{1-v}\int_{\partial\omega} R \neq 0$$

Then the element μ_3 is constructed by solving the Poisson equation on ω:

(II.26)
$$\begin{cases} -\,\Delta\,u_3 = \psi \text{ on } \omega, \\[1mm] u_3 = 0 \text{ on } \partial\omega, \\[1mm] u_3 \in H^1(\omega). \end{cases}$$

Finally we define the operator G from the hyperplane:

(II.27) $\quad K_{f_3} = \left\{ g \,\middle|\, g \in H^{-1/2}(\partial\omega),\ \dfrac{2\,E\,\varepsilon^3}{3(1-v^2)}\langle g, 1\rangle = -\int_\omega f_3 \right\}$

into $H^{1/2}(\partial\omega)$, by:

$$G\,(g) = \psi + \frac{1-v}{R}\frac{\partial\,u_3}{\partial b}$$

where ψ (respectively u_3) is solution to (II.24) (respectively (II.26). From the regularity assumption formulated above, we can deduce that G is perfectly defined from K_{f_3} into

$H^{1/2}(\partial\omega)$. Furthermore,

$$\|G(g)\|_{1/2,\partial\omega} \leq \|\psi\|_{1/2,\partial\omega} + c_0 \left\|\frac{\partial u_3}{\partial b}\right\|_{1/2,\partial\omega}$$

where $c_0 = \sup_{s \in \partial\omega} \left|\frac{1-\nu}{R}\right|(s)$. Then, using the continuity of the trace mapping:

$$\|G(g)\|_{1/2,\partial\omega} \leq c_1\|\psi\|_{1,\omega} + c_0 c_2\|u_3\|_{2,\omega}$$

and finally, from the regularity assumption:

$$\|G(g)\|_{1/2,\partial\omega} \leq c_1\|\psi\|_{1,\omega} + c_3\|\psi\|_{0,\omega} \leq c_4\{\|f_3\|_{0,\omega} + \|g\|_{-1/2,\partial\omega}\} .$$

As G is linear, this estimate proves the continuity of G. Let us set:

$$J(g) = \frac{1}{2}\|G(g)\|_{0,\partial\omega}^2$$

which is a convex and continuous functional defined on K_{f_3}.
Then we introduce the control problem:

$$(\text{II.28}) \quad \begin{cases} \text{minimize } J(g) \\ g \in K_{f_3} \end{cases}$$

We know that there exists a solution g_0 such that $J(g_0) = 0$, because it corresponds to the unique solution to (II.20), (II.21) or else to (II.12), with $g_0 = \dfrac{\partial\psi}{\partial b}$. If there was another solution minimizing J over K_{f_3}, we would have two solutions to (II.12), which is impossible.

The control method presented here is very close to the one suggested almost twenty years ago by R. Glowinski for Stokes problem [4]. The reader is referred to R. Glowinski for further details on this control method (practical aspects: numerical schemes and solution methods).

II.2.4.2 *Clamped plates*

Another important example is $\gamma_0 = \partial\omega$. Then the space V_3^0 is $H_0^2(\omega)$. The variational equation (II.11) is:

(II.29) $\quad$ find $u_3 \in H_0^2(\omega)$ such that :

$$\forall\ v_3 \in H_0^2(\omega),\ \int_\omega \frac{2\,E\,\varepsilon^3}{3\,(1-v^2)}\,\Delta\,u_3\,\Delta\,v_3\ =\ \int_\omega f_3\ v_3$$

Setting for any g in the hyperplane K_{f_3} :

$$-\frac{2\,E\,\varepsilon^3}{3(1-v^2)}\,\Delta\,\psi\ =\ f_3\quad\text{on }\omega$$

$$\frac{\partial\,\psi}{\partial\,b}\ =\ g\quad\text{on }\partial\omega$$

and

$$-\,\Delta\,u_3\ =\ \psi\quad\text{on }\omega$$

$$u_3\ =\ 0\quad\text{on }\partial\omega$$

we define as we did in section II.2.4.1 -case 2-, the control problem (we set $G\,(g)\ =\ \dfrac{\partial\,u_3}{\partial\,b}$ and

we assume the regularity of the Poisson problem (i.e. $u_3 \in H^2(\omega)$ and therefore $\dfrac{\partial u_3}{\partial b} \in L^2(\gamma)$):

$$\text{minimize } J\,(g)\left(=\frac{1}{2}\,\|\,G\,(g)\|^2_{0,\,\partial\omega}\right)$$

$$g\ \in\ K_{f_3}$$

The conclusions are the same concerning the existence and uniqueness of a solution. Here again we refer to R. Glowinski [4], especially in this case because model (II.29) is equivalent to Stokes system.

II.2.4.3 Other boundary conditions

We come back in this section to the general case that we have considered at the beginning of this chapter. Thus the space V_3^0 is defined at (II.7). If we interpret locally (using ad'hoc v_3 test functions), the variational equation (II.12) leads to the following equations:

$$\text{(II.30)} \quad \begin{cases} \dfrac{2\,E\,\varepsilon^3}{3\,(1-v^2)}\,\Delta^2\,u_3 = f_3 \quad \text{on } \omega \\[2em] u_3 = 0 \quad \text{on } \gamma_0 \cup \gamma_1 \\[1.5em] \dfrac{\partial\,u_3}{\partial\,b} = 0 \quad \text{on } \gamma_0 \\[1.5em] \Delta\,u_3 - (1-v)\dfrac{\partial^2\,u_3}{\partial\,s^2} - \dfrac{(1-v)}{R}\dfrac{\partial\,u_3}{\partial\,b} = 0 \quad \text{on } \gamma_1 \cup \gamma_2 \\[2em] \dfrac{\partial}{\partial\,b}(\Delta\,u_3) + (1-v)\dfrac{\partial^3\,u_3}{\partial\,s^2\,\partial\,b} = 0 \quad \text{on } \gamma_2 \end{cases}$$

An interesting work would be to extend the previous boundary conditions to a non-homogeneous loading. For the homogeneous case, let us set:

$$\text{(II.31)} \quad \begin{cases} -\,\Delta\,u_3 = \psi \quad \text{on } \omega \ (\psi \text{ is assumed to be in the space } H^1(\omega)), \\[1em] u_3 = 0 \quad \text{on } \gamma_0 \cup \gamma_1\,, \\[1em] \dfrac{\partial\,u_3}{\partial\,b} = \dfrac{1}{1-v}\int_0^s \int_0^t \dfrac{\partial\,\psi}{\partial\,b}(\xi)\,d\,\xi\,d\,t + A\,s + B \quad \text{on } \gamma_2\,, \\[1em] (\text{where } A \text{ and } B \text{ are two constants on } \gamma_2)\,. \end{cases}$$

Then ψ is defined up to a constant in the space $H^1(\omega)$ by:

$$\text{(II.32)} \quad \begin{cases} \dfrac{-\,2\,E\,\varepsilon^3}{3\,(1-v^2)}\,\Delta\,\psi = f_3 \quad \text{on } \omega \\[1.5em] \dfrac{\partial\,\psi}{\partial\,b} = g \quad \text{on } \gamma_0 \cup \gamma_1 \cup \gamma_2 = \partial\omega \end{cases}$$

where g is an element of the hyperplane K_{f_3}. Then we set:

$$\text{(II.33)} \quad G\,(g) = \begin{cases} \dfrac{\partial\,u_3}{\partial\,b} \quad \text{on } \gamma_0\,, \\[1.5em] \psi - (1-v)\left[\dfrac{\partial^2\,u_3}{\partial\,s^2} + \dfrac{1}{R}\dfrac{\partial\,u_3}{\partial\,b}\right] \quad \text{on } \gamma_1 \cup \gamma_2 \end{cases}$$

and we introduce the control problem:

$$
\text{(II.34)} \quad
\begin{cases}
\text{minimize } \dfrac{1}{2} \, \|G\,(g\,,\,A\,,\,B)\|^{2}_{-\frac{1}{2},\,\partial\omega} \quad (= J\,(g\,,\,A\,,\,B)) \\[2mm]
g \in K_{f_3} \\[2mm]
A\,,\,B \in R
\end{cases}
$$

Here again, $J\,(\ )$ is a convex functional and the minimum (i.e. $J\,(g_0\,,\,A_0\,,\,B_0) = 0$) is obtained

for $g_0 = -\dfrac{\partial}{\partial b}\big(\Delta\,u_3\big),\quad A_0\,s + B_0 = \dfrac{\partial\,u_3}{\partial b} + \dfrac{1}{1-\nu}\displaystyle\int_0^s\int_0^t \dfrac{\partial}{\partial b}\big(\Delta\,u_3\big)\,d\,\xi\,d\,t\,,$ where u_3 is

the unique solution to (II.12) or (II.30). Therefore the solution to (II.34) is also unique. Let us

point out that even if u_3 is in $H^2(\omega)$, then $\dfrac{\partial^2\,u}{\partial\,s^2}$ is only in the space: $H^{1/2}(\omega)$. Therefore it is

necessary to use the norm of the space $H^{-1/2}(\omega)$ in the definition of the criterion G.

II.3   The Reissner-Mindlin-Naghdi model for plates

Let us consider here again a thin plate, the medium surface of which is ω and the thickness 2ε .
Following the idea developped by Ph. Destuynder-Th. Nevers [1], we are going to introduce a
perturbation of the Kirchhoff-Love model which will lead to a plate model similar to the one
suggested by E. Reissner [6] and R. Mindlin [7].   An interesting presentation was also
suggested by P.M. Naghdi [8].   Even if all these formulations start from different
considerations, they lead to mathematical models, the properties of which are identical.   The
only distinction can be found in the coefficients of the transverse shear stiffness.   This
mechanical point which is certainly important is not discussed here.   But the reader can for
instance refer to the papers by P. Ladeveze [9] who suggested an interpretation based on the
boundary layer phenomenon. Concerning the presentation adopted here, the main idea is to
introduce a new variable, say the rotation of the unit normal to the medium surface, as an
independent variable. Then the constraint between this new variable and the derivatives of the
deflection of the medium surface is taken into account through a penalty method.

If the penalty parameter is correctly chosen, then the Reissner-Mindlin or Naghdi model is
obtained by this procedure.   In our presentation the reader should just consider the point as a
remark without any mechanical explanation.   For further details concerning plate and shell
justifications we refer to Ph. Destuynder [10] (concerning Reissner, Mindlin and Naghdi
models).

II.3.1   The penalty method applied to the Kirchhoff-Love model

Let us set:

(II.35) $\theta_\alpha = -\partial_\alpha u_3$ for $\alpha = 1, 2$

where u_3 is solution to the Kirchhoff-Love model given at (II.1) or (II.8). A classical result is that u_3 is also solution to the following optimization problem:

(II.36) $\displaystyle\min_{v_3 \in V_3^0} \frac{1}{2} a(v_3, v_3) - l(v_3)$

where the space V_3^0 is given at (II.7) and the bilinear form $a(.,.)$ and the linear form $l(.)$ are explicited at (II.9). Then, noticing relation (II.35), one can write :

(II.37)
$$
\begin{cases}
a(u_3, v_3) \overset{\text{definition}}{\equiv} k(\theta, \mu) = \displaystyle\int_\omega \frac{2E\varepsilon^3}{3(1-v^2)} \left\{ (1-v)\,\gamma_{\alpha\beta}(\theta)\gamma_{\alpha\beta}(\mu) + v\gamma_{\lambda\lambda}(\theta)\gamma_{\xi\xi}(\mu) \right\}. \\[2mm]
\theta_\alpha = -\partial_\alpha u_3 \quad \text{and} \quad \mu_\alpha = -\partial_\alpha v_3
\end{cases}
$$

where we recall that : $\gamma_{\alpha\beta}(\mu) = \frac{1}{2}(\partial_\alpha\mu_\beta + \partial_\beta\mu_\alpha)$. Thus problem (II.36) can be formulated as follows:

$$
\begin{cases}
\min \dfrac{1}{2} k(\mu, \mu) - l(v_3) \\[2mm]
v_3 \in V_3^0 \\[2mm]
\mu_\alpha = -\partial_\alpha v_3, \quad \alpha = 1, 2
\end{cases}
$$

The relation $\mu_\alpha + \partial_\alpha v_3 = 0$ can be considered as a constraint.

First of all let us make explicit the functional space to which μ_α belongs. We set (the index "t" means: tangential; the vector θ being tangent to the surface ω):

$$
W_t = \left\{ \mu \mid \mu = (\mu_\alpha) \in (H^1(\omega))^2,\ \mu_\alpha = 0 \text{ on } \gamma_0,\ \mu_s = \mu_\alpha a_\alpha = 0 \text{ on } \gamma_1 \right\}
$$

It appears clearly that if $v_3 \in V_3^0$ then $\mu = (\mu_\alpha)$ where $\mu_\alpha = -\partial_\alpha v_3$, belongs to W_t. Let us then introduce a penalty method for (II.37); for any small parameter – say η – we set:

(II.38)

$$\left\{ \begin{array}{l} \text{minimize } \frac{1}{2} k\left(\mu, \mu\right) - l\left(v_3\right) + \frac{1}{2\eta} \int_\omega \sum_{\alpha=1,2} \left(\mu_\alpha + \partial_\alpha v_3\right)^2 \\[2mm] v_3 \in V_3 \\[1mm] \mu \in W_t \end{array} \right.$$

where :

(II.39)

$$\left\{ \begin{array}{l} V_3 = \left\{ v \mid v \in H^1(\omega), \ v = 0 \quad \text{on } \gamma_0 \cup \gamma_1 \right\} \\[2mm] W_t = \left\{ \mu \mid \mu = (\mu_\alpha) \in \left(H^1(\omega)\right)^2, \ \mu_\alpha = 0 \quad \text{on } \gamma_0, \ \mu_s = \mu_\alpha a_\alpha = 0 \quad \text{on } \gamma_1 \right\} \end{array} \right.$$

Remark II.3

The way we introduced the penalty term is rather crude. We mean that nothing guarantees that (II.38) will be a nice approximation of the Kirchhoff-Love model. As a matter of fact it is proved later on that the solution to (II.38) – say $\left(u_3^\eta, \theta^\eta\right)$ – is close to $\left(u_3, -\partial_\alpha u_3\right)$ as soon as η is small enough. But a different approximated model for bending plate– that we think to be more appropriate – is presented in section II.3.2.   ∎

Remark II.4

Because (II.38) is the minimization of a positive quadratic form, it is equivalent to the variational equation:

(II.40)

$$\left\{ \begin{array}{l} \text{find } \left(\theta^\eta, u_3^\eta\right) \in W_t \times V_3 \text{ such that :} \\[3mm] \forall \ \mu \in W_t, \ k\left(\theta^\eta, \mu\right) + \frac{1}{\eta} \int_\omega \left(\theta_\alpha^\eta + \partial_\alpha u_3^\eta\right) \mu_\alpha = 0 \\[3mm] \forall \ v_3 \in V_3, \ \frac{1}{\eta} \int_\omega \left(\theta_\alpha^\eta + \partial_\alpha u_3^\eta\right) \partial_\alpha v_3 = \int_\omega f_3 \ v_3 \end{array} \right.$$

   ∎

When choosing $\eta = \dfrac{(1+v)}{\varepsilon E}$, the above model is the one suggested – for instance – by P.M. Naghdi, ε being half the thickness of the plate. With another choice for η we can derive the Reissner or Mindlin model. But this has no influence on the mathematical results obtained for η tends to zero. Furthermore it is interesting to introduce the "pseudo" transverse shear stress by:

(II.41) $q_\alpha^\eta = \frac{1}{\eta}\left(\theta_\alpha^\eta + \partial_\alpha u_3^\eta\right)$ for $\alpha = 1, 2$

Then (II.40) is equivalent to (this is rigorous!):

(II.42)
$$
\left\{
\begin{array}{l}
\text{find } \left(\theta^\eta, u_3^\eta, q^\eta\right) \in W_t \times V_3 \times H_t \text{ such that :} \\[2em]
\forall \; \mu \in W_t, \; k\left(\theta^\eta, \mu\right) + \displaystyle\int_\omega q_\alpha^\eta \mu_\alpha = 0, \\[2em]
\forall \; v_3 \in V_3, \; \displaystyle\int_\omega q_\alpha^\eta \partial_\alpha v_3 = \int_\omega f_3 v_3, \\[2em]
\forall \; p \in H_t, \; \eta \displaystyle\int_\omega q_\alpha^\eta p_\alpha = \int_\omega p_\alpha \left(\theta_\alpha^\eta + \partial_\alpha u_3^\eta\right),
\end{array}
\right.
$$

where we have set

(II.43) $H_t = \left\{ p \mid p = (p_\alpha) \in \left[L^2(\omega)\right]^2 \right\}$

(here again the index "t" means tangential because the transverse shear has only tangential components). One can observe that q^η does not have the dimensions of a stress term. It depends on the choice of η. This is why we used the terminology "pseudo". The expression of the transverse shear stress is recalled in chapter I for the Kirchhoff-Love model.

II.3.1.1 Existence and uniqueness of a solution

Let us first state the result that we are going to prove.

Theorem II.1.
Let us assume that the element f_3 is an element of the space $L^2(\omega)$. Furthermore the boundary of ω is supposed to be smooth enough (piecewise $W^{1,\infty}$ is sufficient). Then the variational system (II.40), (or II.42) has a unique solution.   ∎

Proof of Theorem II.1.
For sake of clarity, let us introduce the notation:

$$X^\eta = \left(\theta^\eta, u_3^\eta\right) \qquad Y = (\mu, v_3)$$

and the space, which is obviously an Hilbert space :

$$\vartheta = W_t \times V_3 .$$

Finally the bilinear form $a^\eta (. , .)$ is defined on $\vartheta \times \vartheta$ by:

$$a^\eta (X, Y) = k\left(\theta, \mu\right) + \frac{1}{\eta} \int_\omega \left(\theta_\alpha + \partial_\alpha u_3\right)\left(\mu_\alpha + \partial_\alpha v_3\right)$$

and the linear form:

$$l (Y) = \int_\omega f_3 \, v_3$$

Then (II.40) can also be written:

$$\text{(II.44)} \quad \begin{cases} \text{find } X^\eta \in \vartheta \text{ such that :} \\[2mm] \forall \ Y \in \vartheta, a^\eta \left(X^\eta, Y\right) = l (Y) \end{cases}$$

In order to prove Theorem II.1, we plan to apply Lax-Milgram Theorem [1]. First of all, let us notice that for any $Y \in \vartheta$ and for any given positive number K (smaller than $\frac{1}{\eta_0}$), one has for $\eta < \eta_0$:

$$a^\eta (Y, Y) \geq k (\mu, \mu) + \sum_{\alpha, \beta = 1, 2} K \left\| \mu_\alpha + \partial_\alpha v_3 \right\|_{0, \omega}^2$$

But $k (. , .)$ is such that

$$k (\mu, \mu) \geq \frac{2 E \varepsilon^3}{3 (1 + \nu)} \sum_{\alpha, \beta = 1, 2} \left\| \gamma_{\alpha \beta} (\mu) \right\|_{0, \omega}^2$$

and from Korn inequality and its consequences, assuming for instance that $\text{meas}(\gamma_0) \neq 0$ (see for instance G. Duvaut -J. L. Lions [12]):

$$k (\mu, \mu) \geq c \sum_{\alpha = 1, 2} \left\| \mu_\alpha \right\|_{1, \omega}^2$$

where c is a constant, or else:

$$\forall \ \mu \in \ W_t \ , \ k\left(\mu \ , \mu\right) \ \geq \ c \left\|\mu\right\|_{W_t}^2$$

(the norm on W_t is obviously the one induced by the Sobolev space $H^1\left(\omega\right)$). Then one has also:

$$\sum_{\alpha = 1, 2} \left\|\mu_\alpha + \partial_\alpha \ v_3\right\|_{0, \omega}^2 \ = \ \sum_{\alpha = 1, 2} \left(\left\|\mu_\alpha\right\|_{0, \omega}^2 + \left\|\partial_\alpha \ v_3\right\|_{0, \omega}^2\right) \ + \ 2 \int_\omega \mu_\alpha \ \partial_\alpha \ v_3$$

and from the inequality:

$$\forall \ \xi \ > \ 0 \ , \ 2 \left|a \ b\right| \ \leq \ \xi \ a^2 \ + \ \frac{1}{\xi} \ b^2,$$

we deduce that for any strictly positive number ξ :

$$\sum_{\alpha = 1, 2} \left\|\mu_\alpha + \partial_\alpha \ v_3\right\|_{0, \omega}^2 \ \geq \ \sum_{\alpha = 1, 2} \left\|\mu_\alpha\right\|_{0, \omega}^2 \left(1 - \xi\right) + \left\|\partial_\alpha \ v_3\right\|_{0, \omega}^2 \left(1 - \frac{1}{\xi}\right) .$$

Hence noticing that for $\left(1 - \xi\right) \ < \ 0$ (or else $\xi > 1$) and because the norm in the space $(L^2(\omega))^2$ is smaller than the one in the space W_t, one has (c_0 being a positive constant):

$$\sum_{\alpha = 1, 2} \left\|\mu_\alpha + \partial_\alpha \ v_3\right\|_{0, \omega}^2 \ \geq \ c_0 \left(1 - \frac{1}{\xi}\right) \left\|v_3\right\|_{1, \omega}^2 \ - \ \left(\xi - 1\right) \left\|\mu\right\|_{W_t}^2$$

because $v_3 \ \rightarrow \ \left[\sum_{\alpha = 1, 2} \left\|\partial_\alpha \ v_3\right\|_{0, \omega}^2\right]^{1/2}$ is a norm on V_3 which is equivalent to the one of $H^1\left(\omega\right)$. This is a classical result named: " Generalized Poincaré inequality". Finally for any $\xi > 1$ we proved that:

$$a^\eta \left(Y \ , Y\right) \ \geq \ \left[c - K\left(\xi - 1\right)\right] \left\|\mu\right\|_{W_t}^2 \ + \ c_0 \left(1 - \frac{1}{\xi}\right) K \left\|v_3\right\|_{1, \omega}^2$$

(let us underline that for $\xi > 1$ then $1 - \frac{1}{\xi} \ > \ 0$). Thus, if we choose:

$$0 \ < \ K \ < \min\left(\frac{c}{\xi - 1} \ , \frac{1}{\eta_0}\right)$$

we obtain that there exists a strictly positive constant, say c_1 , such that:

(II.45) $\forall \ Y \in \vartheta$, $a^\eta (Y, Y) \geq c_1 \left(\|\mu\|_{W_t}^2 + \|v_3\|_{1, \omega}^2 \right) = c_1 \|Y\|_\vartheta^2$.

This is the ellipticity of $a^\eta (. , .)$ over ϑ. One can notice that this ellipticity is uniform with respect to η as soon as it is small enough. With another respect, the continuity of $a^\eta (. , .)$ and $l (.)$, on the space ϑ are quite obvious. Hence we can make use of Lax-Milgram Theorem [1]. The existence (and uniqueness) of q^η in the space H_t (see (II.43) for the definition of H_t) is a consequence of the relation:

$$\eta \ q_\alpha^\eta = \theta_\alpha^\eta + \partial_\alpha u_3^\eta , \quad \alpha = 1, 2 \qquad\qquad \blacksquare$$

II.3.1.2 *Convergence of* $\left(\theta^\eta, u_3^\eta \right)$

We use a classical procedure based on an a priori estimate.

Theorem II.2
Let us assume for instance that $f_3 \in L^2 (\omega)$. *Then the solution* $\left(\theta^\eta, u_3^\eta \right)$ *to (II.42) is such that:*

$$\begin{cases} \lim_{\eta \to 0} \theta_\alpha^\eta = \theta_\alpha & \text{in } L^2 (\omega) \text{ strong,} \\[2ex] \lim_{\eta \to 0} u_3^\eta = u_3 & \text{in } H^1 (\omega) \text{ strong,} \end{cases}$$

where u_3 *is the solution to Kirchhoff-Love model and* $\theta_\alpha = - \partial_\alpha u_3$.

Remark II.5
It seems quite impossible to improve the Theorem II.2. This is due to the free boundary (γ_2). As a matter of fact, the boundary conditions on γ_2 are different for Kirchhoff-Love or Reissner-Mindlin-Naghdi models. In Kirchhoff-Love model, one has (see chapter 1) on the free edge γ_2:

$$\begin{cases} m_{\alpha\beta} \ b_\alpha \ b_\beta = 0 , \\[1ex] \text{and} \\[1ex] \partial_s \left(m_{\alpha\beta} \ a_\alpha \ b_\beta \right) + \partial_\alpha \ m_{\alpha\beta} \ b_\beta = 0 \end{cases}$$

and for Reissner-Mindlin-Naghdi model, one has:

$$\begin{cases} m_{\alpha\beta}\, b_\alpha\, b_\beta = 0, \\ \text{and} \\ m_{\alpha\beta}\, a_\alpha\, b_\beta = 0. \end{cases}$$

Hence we cannot hope for a convergence result in a functional space which gives a meaning to the boundary condition on γ_2. This is the reason why we claim that Theorem II.2 is difficult to improve. We shall discuss this point in section II.3.2 where an "improvement" of Reissner-Mindlin-Naghdi model is suggested. ∎

Proof of Theorem II.2

Let us start with an a priori estimate on $\left(\theta^\eta, u_3^\eta\right)$. Choosing $\mu = \theta^\eta$ and $v_3 = u_3^\eta$ in (II.40), one obtains:

$$(II.46)\quad k\left(\theta^\eta, \theta^\eta\right) + \frac{1}{\eta} \sum_{\alpha=1,2} \left\| \theta_\alpha^\eta + \partial_\alpha u_3^\eta \right\|_{0,\omega}^2 \leq \| f_3 \|_{0,\omega}\, \| u_3^\eta \|_{0,\omega}$$

or else using the ellipticity of $a^\eta(.\,,.)$ established in the proof of Theorem II.1, and which is independent of η (as soon as η is small enough; see (II.45)):

$$c\left(\| \theta^\eta \|_{W_t}^2 + \| u_3^\eta \|_{1,\omega}^2 \right) \leq \| f_3 \|_{0,\omega}\, \| u_3^\eta \|_{0,\omega}$$

from which we deduce for any real number $\xi > 0$:

$$c\| \theta^\eta \|_{W_t}^2 + \left(c - \frac{\xi}{2} \right) \| u_3^\eta \|_{1,\omega}^2 \leq \frac{1}{2\xi} \| f_3 \|_{0,\omega}^2$$

Hence choosing $\xi < 2\,c$ we deduce the a priori estimate:

$$(II.47)\quad \forall\ \eta\ :\ \| \theta^\eta \|_{W_t}^2 + \| u_3^\eta \|_{1,\omega}^2 \leq cte\ .$$

In addition, from (II.46), we deduce that:

$$(II.48)\quad \forall\ \eta\ :\ \sum_{\alpha=1,2} \left\| \theta_\alpha^\eta + \partial_\alpha u_3^\eta \right\|_{0,\omega}^2 \leq \eta\ cte$$

The bounded convex sets of an Hilbert space are weakly compact. Furthermore the natural embedding from $H^1(\omega)$ into $L^2(\omega)$ is compact. Hence there exists a subsequence of $\left(\theta^\eta, u_3^\eta\right)$, denoted by $\left(\theta^{\eta'}, u_3^{\eta'}\right)$ and such that:

$$(\text{II.49}) \quad \left| \begin{array}{l} \lim_{\eta' \to 0} \theta_\alpha^{\eta'} = \theta_\alpha^* \quad \text{in} \quad H^1(\omega) \text{ weak}, \\[2mm] \lim_{\eta' \to 0} \theta_\alpha^{\eta'} = \theta_\alpha^* \quad \text{in} \quad L^2(\omega) \text{ strong}, \\[2mm] \lim_{\eta' \to 0} u_3^{\eta'} = u_3^* \quad \text{in} \quad H^1(\omega) \text{ weak}, \\[2mm] \lim_{\eta' \to 0} u_3^{\eta'} = u_3^* \quad \text{in} \quad L^2(\omega) \text{ strong}. \end{array} \right.$$

one has:

$$\sum_{\alpha=1,2} \left\| \theta_\alpha^{\eta'} + \partial_\alpha u_3^{\eta'} \right\|_{0,\omega}^2 = \sum_{\alpha=1,2} \left(\left\| \theta_\alpha^{\eta'} \right\|_{0,\omega}^2 + 2 \int_\omega \theta_\alpha^{\eta'} \partial_\alpha u_3^{\eta'} + \left\| u_\alpha^{\eta'} \right\|_{1,\omega}^2 \right)$$

(we used the norm on V_3 induced by the one of the derivatives in $L^2(\omega)$), and therefore from (II.48) and (II.49):

$$\lim_{\eta' \to 0} \left\| u_3^{\eta'} \right\|_{V_3}^2 = - \sum_{\alpha=1,2} \left\| \theta_\alpha^* + \partial_\alpha u_3^* \right\|_{0,\omega}^2 + \left\| u_3^* \right\|_{V_3}^2$$

But from the semi-lower continuity of a continuous convex function for the weak topology:

$$0 \le \left\| \theta_\alpha^* + \partial_\alpha u_3^* \right\|_{0,\omega} \le \lim_{\eta' \to 0} \inf \left\| \theta_\alpha^{\eta'} + \partial_\alpha u_3^{\eta'} \right\|_{0,\omega} = 0$$

we deduce that:

$$\lim_{\eta' \to 0} \left\| u_3^{\eta'} \right\|_{V_3}^2 = \left\| u_3^* \right\|_{V_3}^2$$

Finally (this is a classical result):

$$\lim_{\eta' \to 0} \left\| u_3^{\eta'} - u_3^* \right\|_{V_3}^2 = \lim_{\eta' \to 0} \left[\left\| u_3^{\eta'} \right\|_{V_3}^2 - 2 \int_\omega \partial_\alpha u_3^{\eta'} \partial_\alpha u_3^* \right] + \left\| u_3^* \right\|_{V_3}^2 = 0,$$

which proves the strong convergence of $u_3^{\eta'}$ to u_3^* in $H^1(\omega)$, because of the equivalence between the norms in the spaces V_3 and $H^1(\omega)$. Furthermore we proved in the above expressions that:

$$\theta_\alpha^* + \partial_\alpha u_3^* = 0 \quad \text{on} \quad \omega,$$

which implies (because $\theta^* \in W_t$):

$$u_3^* \in V_3^0 \quad \text{(see definition (II.7)} \quad .$$

Let us now come back to the equation (II.40) which characterizes the solution $\left(\theta^{\eta'}, u_3^{\eta'}\right)$. Then choosing $v_3 \in V_3^0$ and $\mu_\alpha = -\partial_\alpha v_3$ (hence $\mu \in W_t$!), we obtain:

$$\forall \ v_3 \in V_3^0, \ k\left(\theta^{\eta'}, \mu\right) = \int_\omega f_3 \ v_3$$

Using the weak convergence of θ^η to θ^*, we get:

$$\left| \begin{array}{l} \theta_\alpha^* = -\partial_\alpha u_3^* \in W_t, u_3^* \in V_3^0 \quad \text{and}: \\[2mm] \forall \ v_3 \in V_3^0, \ k\left(\theta^*, \mu\right) = \int_\omega f_3 \ v_3 . \qquad \left(\mu_\alpha = -\partial_\alpha v_3\right) \end{array} \right.$$

and from (II.37),

$$\left| \begin{array}{l} u_3^* \in V_3^0 \\[2mm] \forall \ v_3 \in V_3^0, \ a\left(u_3^*, v_3\right) = \int_\omega f_3 \ v_3 , \end{array} \right.$$

which proves that $u_3^* = u_3$ is the solution to the Kirchhoff-Love model (see (II.36) for instance). Because of the uniqueness of u_3, we can conclude that the whole sequence u_3^η tends to u_3 (idem for θ_α^η which tends to : $\theta_\alpha = -\partial_\alpha u_3$).   ∎

Remark II.6

It is possible to derive from Theorem II.2 an indication on the convergence of the transverse shear stress. Let us consider the first relation (II.42) and let us restrict μ to the space $\left(H_0^1(\omega)\right)^2$ which is obviously a closed subspace of W_t. . Then we get:

$$\forall \mu \in \left(H_0^1(\omega)\right)^2, \ \int_\omega q_\alpha^\eta \mu_\alpha = -k\left(\theta^\eta, \mu\right) \quad .$$

But from the definition of the $H^{-1}(\omega)$ norm, one has:

$$\sup_{\mu = (\mu_\alpha) \in \left(H_0^1(\omega)\right)^2} \frac{\displaystyle\int_\omega q_\alpha^\eta \, \mu_\alpha}{\|\mu\|_{w_t}} = \sum_{\alpha = 1, 2} \|q_\alpha^\eta\|_{-1, \omega}$$

and we deduce that there exists a strictly positive constant such that:

$$\sum_{\alpha = 1, 2} \|q_\alpha^\eta\|_{-1, \omega} \leq c \|\theta^\eta\|_{w_t} \leq cte.$$

Therefore there exists a subsequence, say $q_\alpha^{\eta'}$, such that (weak convergence in $H^{-1}(\omega)$):

$$\forall \mu \in \left(H_0^1(\omega)\right)^2, \quad \lim_{\eta' \to 0} \int_\omega q_\alpha^{\eta'} \mu_\alpha = \langle q_\alpha^*, \mu_\alpha \rangle$$

where $q_\alpha^* \in H^{-1}(\omega)$ and $<,>$ denotes the duality between $H^{-1}(\omega)$ and $H_0^1(\omega)$. Hence:

(II.49) $\forall \mu \in \left(H_0^1(\omega)\right)^2, \quad <q_\alpha^*, \mu_\alpha> + k(\theta, \mu) = 0$

θ being the limit of all the subsequences of θ^η. It is easy to check that q_α^* is unique because θ is unique. Hence the whole sequence q_α^η tends to q_α^* in $H^{-1}(\omega)$ weakly.

Let us complete this remark by expliciting the term q_α^*. From (II.49) we deduce:

(II.50) $q_\alpha^* = \partial_\beta m_{\alpha\beta} = q_\alpha$ (which a distribution derivative, see chapter I)

where we have set:

$$m_{\alpha\beta} = \frac{-2 E \varepsilon^3}{3(1 - v^2)} \left\{ (1 - v) \partial_{\alpha\beta} u_3 + v \Delta u_3 \, \delta_{\alpha\beta} \right\}.$$

■

II.3.2   A correction to the penalty method

The idea which is presented in this section is due to Ph. Destuynder and Th. Nevers [1] and aims at improving the convergence of $\left(\theta^\eta, u_3^\eta, q^\eta\right)$ to $\left(\theta, u_3, q\right)$ as η tends to zero. The last triplet is solution to Kirchhoff-Love plate model. The main difficulty that we met was due to the difference between the boundary condition satisfied by the bending moments on the free edge, between Kirchhoff-Love and the penalty models. Obviously, as it appears in the Theorems formulated hereafter, when there is no free edge, there is no problem and the error estimates given in this section with the modified penalty model can also be applied to the previous model explicited in section II.3.1. The basic idea is to take into account the constraint:

$$\theta_s + \partial_s u_3 = 0 \quad \text{on } \gamma_2$$

in the penalty method. One could object that it is not necessary because the relation :

$$\theta_\alpha + \partial_\alpha u_3 = 0 \quad \text{on } \omega \quad \text{for } \alpha = 1, 2$$

has already been taken into account through the penalty term :

$$\frac{1}{2\eta} \sum_{\alpha = 1, 2} \left\| \theta_\alpha^\eta + \partial_\alpha u_3^\eta \right\|_{0, \omega}^2 .$$

Unfortunately this is only meaningful in the space $L^2(\omega)$ for which the trace of $\theta_\alpha^\eta + \partial_\alpha u_3^\eta$ on the boundary of ω does not make sense. Even if $\theta_\alpha^\eta \in H^1(\omega)$ and $u_3^\eta \in H^1(\omega)$ have a trace on γ_2 they are not dependent. For instance from

$$\theta_s = \theta_\alpha a_\alpha = - a_\alpha \partial_\alpha u_3 = - \partial_s u_3 \quad \text{on } \gamma_2$$

which is satisfied by the Kirchhoff-Love fields we deduce the boundary condition (see section II.2):

$$\partial_s \left(m_{\alpha\beta} a_\alpha b_\beta \right) + \partial_\alpha m_{\alpha\beta} b_\beta = 0 \quad \text{and} \quad m_{\alpha\beta} b_\alpha b_\beta = 0 \quad \text{on } \gamma_2 .$$

On the contrary, in the penalty model, θ_s^η and $\partial_s u_3^\eta$ are independent of γ_2, which leads to the boundary condition:

$$m_{\alpha\beta}^\eta a_\alpha b_\beta = 0 , m_{\alpha\beta}^\eta b_\alpha b_\beta = 0, \text{ and } \partial_\alpha m_{\alpha\beta}^\eta b_\beta = 0 \quad \text{on } \gamma_2$$

($m_{\alpha\beta}^\eta$ are the bending moments issued from the penalty model). Hence our goal is to modify this last boundary conditions so that both the Kirchhoff-Love and the penalty model have almost the same behaviour near the free edge γ_2 . In a first step we proceed in a formal way using a Lagrangian technique. The mathematical justification is then given.

II.3.2.1 A formal derivation of the modified penalty method.

Let us recall that the penalty model consists in finding $\left(\theta^{\eta}, u_3^{\eta}\right) \in W_t \times V_3$ which minimizes the quantity:

$$\frac{1}{2}\, k\,(\mu\,,\mu) + \frac{1}{2\,\eta} \sum_{\alpha=1,2} \left\| \mu_\alpha + \partial_\alpha v_3 \right\|_{0,\omega}^2 - \int_\omega f_3\, v_3$$

for $(\mu\,,v_3) \in W_t \times V_3$. The new idea is to take into account (exactly!) the condition

$$\mu_s + \partial_s v_3 = 0 \quad \text{on } \gamma_2\,.$$

Therefore we restrict the space $\vartheta \in W_t \times V_3$ to a subspace – say ϑ_0 – defined by:

(II.51)
$$\begin{aligned}\vartheta_0 = \Big\{ (\mu\,,v_3) \,\big|\, \mu \in \left(H^1(\omega)\right)^2,\ v_3 \in H^1(\omega),\ \mu = 0 \ \text{on } \gamma_0\,. \\ v_3 = 0 \ \text{on } \gamma_0 \cup \gamma_1,\ \mu_s = \mu_\alpha\, a_\alpha = 0 \ \text{on } \gamma_1\,, \\ \text{and } \mu_s + \partial_s\, v_3 = 0 \ \text{on } \gamma_2 \Big\}\,.\end{aligned}$$

As a matter of fact, the relation

$$\mu_s + \partial_s\, v_3 = 0 \quad \text{on } \gamma_2$$

defines a closed subspace of ϑ, because on the one hand of the continuity of trace from $H^1(\omega)$ into $H_{00}^{1/2}(\partial\omega)$ and on the other hand because of the continuity of the derivation from $H_{00}^{1/2}(\partial\omega)$ into $H^{-1/2}(\partial\omega)$. At this point it is worth noticing that the previous constraint is meaningful in the Sobolev space $H^{-1/2}(\gamma_2)$. The modified penalty model is thus defined as follows:

(II.52)
$$\left\{ \begin{aligned} &\text{find}\left(\theta^{\eta}, u_3^{\eta}\right) \in \vartheta_0 \ \text{which minimizes:} \\[1mm] &(\mu,v_3) \in \vartheta_0 \ \rightarrow \frac{1}{2}\, k\,(\mu\,,\mu) + \frac{1}{2\,\eta} \sum_{\alpha=1,2} \left\| \mu_\alpha + \partial_\alpha v_3 \right\|_{0,\omega}^2 - \int_\omega f_3\, v_3 \end{aligned} \right.$$

Obviously the constraint on γ_2 which is included in the definition of ϑ_0 is not very convenient. This is the reason why we prefer to get rid of it, using a duality method (which is justified in section II.3.2.2).

The space $H_{00}^{1/2}(\gamma_2)$ is equal to $H^{1/2}(\gamma_2)$ if γ_2 is a closed boundary. The difference occurs only in the local behaviour of the functions near the edge of γ_2 if it is not closed (see figure II.2). We also use the space $H_{00}^{-1/2}(\gamma_2)$ which is the dual space of $H_{00}^{1/2}(\gamma_2)$. Another functional space which is important for our purpose is denoted by: $H_+^{1/2}(\gamma_2)$ and is defined by:

$$H_+^{1/2}(\gamma_2) = \left\{ t \mid t \in H^{-1/2}(\gamma_2) , \; \partial_s t \in H^{-1/2}(\gamma_2) \right\}.$$

It is equipped with the norm of the space $H^{1/2}(\gamma_2)$ Obviously this space is identical to $H^{1/2}(\gamma_2)$ when the boundary γ_2 is closed. In order to make the notations simple, we use the expression: $\|\cdot\|_{1/2,\gamma_2}$ for the norms of the spaces $H_{00}^{1/2}(\gamma_2)$, $H_+^{1/2}(\gamma_2)$ or $H^{1/2}(\gamma_2)$ and $\|\cdot\|_{-1/2,\gamma_2}$ for the spaces : $H_{00}^{-1/2}(\gamma_2)$, and $H^{-1/2}(\gamma_2)$. As a matter of fact, it seems that no confusion is possible. Let us now introduce the duality formulation for the constraint contained in the definition of the space ϑ_0. Thus we introduce a Lagrange multiplier – say r – which belongs to the space $H_+^{1/2}(\gamma_2)$, to the constraint :

$$\theta_s^\eta + \partial_s u_3^\eta = 0 \qquad \text{on } \gamma_2 .$$

The choice of the space $H_+^{1/2}(\gamma_2)$ is done so that the following Lagrangian is defined (i.e. is calculable):

(II.53)
$$\left\{ L(Y, P) = \frac{1}{2} k(\mu , \mu) + \langle \mu_s + \partial_s v_3 , r \rangle + \int_\omega p_\alpha (\mu_\alpha + \partial_\alpha v_3) - \frac{\eta}{2} \int_\omega p_\alpha\, p_\alpha - l(v_3) \right.$$

where:

$$\left(\vartheta = W_t \times V_3 \quad \text{and} \quad M = H_t \times H_+^{1/2}(\gamma_2) \right) \quad Y = (\mu , v_3) \in \vartheta , \quad P = (p , r) \in M :$$

and the problem that we consider consists in finding an element $(X^\eta , Q^\eta) \in \vartheta \times M$ such that:

(II.54)
$$\left\{ \begin{array}{l} \forall \; Y \in \vartheta , \forall \; P \in M , L(X^\eta , P) \le L(X^\eta , Q^\eta) \le L(Y , Q^\eta) \\ \text{with } X^\eta = (\theta^\eta , u_3^\eta), \; Q^\eta = (q^\eta , t^\eta) \end{array} \right.$$

At the present time, nothing guarantees that (II.54) and (II.52) are equivalent. This is proved in II.3.2.2 as we already mentioned. But let us first interpret an assumed existing solution to (II.54) just by considering the optimality relationships deduced from (II.54):

(II.55)

$$
\begin{cases}
\forall\ \mu \in W_t\,,\ k\left(\theta^\eta\,,\mu\right) + \displaystyle\int_\omega q_\alpha^\eta\,\mu_\alpha + \left\langle\mu_s\,,t^\eta\right\rangle = 0\,, \\[2ex]
\forall\ v_3 \in V_3\,,\ \displaystyle\int_\omega q_\alpha^\eta\,\partial_\alpha\,v_3 + \left\langle\partial_s\,v_3\,,t^\eta\right\rangle = \displaystyle\int_\omega f_3\,v_3\,, \\[2ex]
\forall\ p \in H_t\,,\eta\ \displaystyle\int_\omega q_\alpha^\eta\,p_\alpha = \displaystyle\int_\omega p_\alpha\left(\theta_\alpha^\eta + \partial_\alpha\,u_3^\eta\right), \\[2ex]
\forall\ r \in H_+^{1/2}\,(\gamma_2),\ \left\langle\theta_s^\eta + \partial_s\,u_3^\eta\,,r\right\rangle = 0\,.
\end{cases}
$$

where the notation $<\,,>$ stands for the duality between $H_{00}^{1/2}\,(\gamma_2)$ and $H_{00}^{-1/2}\,(\gamma_2)$.

Figure II.2

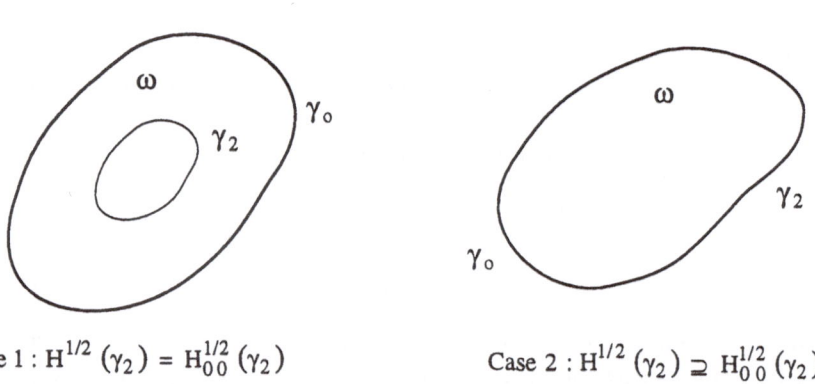

Case 1 : $H^{1/2}\,(\gamma_2) = H_{00}^{1/2}\,(\gamma_2)$ Case 2 : $H^{1/2}\,(\gamma_2) \supsetneq H_{00}^{1/2}\,(\gamma_2)$

The local equations corresponding to (II.55) are obtained by integration by parts. Thus we obtain:

Constitutive relationships

$$m^\eta_{\alpha\beta} = \frac{2E\varepsilon^3}{3(1-v^2)} \left\{ (1-v)\gamma_{\alpha\beta}(\theta^\eta) + v\,\gamma_{\lambda\lambda}(\theta^\eta)\,\delta_{\alpha\beta} \right\},$$

$$\gamma_{\alpha\beta}(\theta^\eta) = \frac{1}{2}\left(\partial_\alpha\theta^\eta_\beta + \partial_\beta\theta^\eta_\alpha \right), \qquad \eta\, q^\eta_\alpha = \theta^\eta_\alpha + \partial_\alpha u^\eta_3 \quad \text{on } \omega$$

Local equilibrium equations

$$\partial_\beta m^\eta_{\alpha\beta} = q^\eta_\alpha\,,$$

$$\partial_\alpha q^\eta_\alpha + f_3 = 0 \quad \text{on } \omega\,,$$

Boundary conditions

(II56)

$$\theta^\eta_\alpha = 0 \quad \text{on } \gamma_0\,, \quad \theta^\eta_s = \theta^\eta_\alpha a_\alpha = 0 \quad \text{on } \gamma_1$$

$$u^\eta_3 = 0 \quad \text{on } \gamma_0 \cup \gamma_1$$

$$m^\eta_{\alpha\beta}\, b_\alpha\, b_\beta = 0 \quad \text{on } \gamma_1 \cup \gamma_2$$

$$m^\eta_{\alpha\beta}\, b_\alpha\, a_\beta + t^\eta = 0 \quad \text{on } \gamma_2$$

$$q^\eta_\alpha\, b_\alpha - \partial_s t^\eta = 0 \quad \text{on } \gamma_2$$

The last two equations imply:

$$\partial_s\left(m^\eta_{\alpha\beta}\, a_\beta\, b_\alpha \right) + \partial_\alpha m^\eta_{\alpha\beta}\, b_\alpha = 0 \quad \text{on } \gamma_2$$

which is precisely one of the Kirchhoff-Love boundary conditions on a free edge (one of the two, as a matter of fact).

II.3.2.2 Existence and uniqueness of a solution

The goal of this section is both to prove the existence and uniqueness of a solution to (II.55) and the equivalence with (II.52). First of all let us consider that (II.55) has a unique solution say: θ^η, u^η, q^η and t^η. The simplest way to prove that θ^η, u^η is solution to (II.52), is certainly to eliminate q^η and t^η from (II.55). This is straightforwardly obtained by limiting the test functions (μ,v) in (II.55) to the space ϑ_0. Thus we get that a solution to (II.55) is also a solution to optimality conditions of (II.52). Then the ellipticity of the bilinear form a^η (. , .) on the space ϑ which was proved in section II.3.1.1 (inequality (II.45)) can also be applied on the closed subspace ϑ_0 of ϑ. The proof of the existence and uniqueness of a solution to (II.52) is therefore quite obvious from the results of section II.3.1.1 (Theorem II.1). The job is still to be done for the existence and uniqueness of q^η and t^η in order to satisfy the variational equations (II.55) (θ^η and u_3^η being now characterized as the unique solution to (II.52)). As a matter of fact the third equation (II.55) is equivalent to:

$$\eta \, q_\alpha^\eta \;=\; \theta_\alpha^\eta + \partial_\alpha u_3^\eta \qquad \alpha = 1\,,2 \ \ \text{on} \ \omega$$

which perfectly defines q^η as an element of the space H_t (as soon as $\eta \neq 0$).

The last relation (II.55) is satisfied by θ^η and u_3^η because $\left(\theta^\eta\,,\,u_3^\eta\right) \in \vartheta_0$. Finally, the only thing to be proved is the existence and uniqueness of an element t^η in the space $H_+^{1/2}(\gamma_2)$ such that the first two relations (II.55) are satisfied. One way to obtain this result is the Babuska Lemma [15], which can be stated as follows :

Lemma II.1:
Let ϑ_1 and ϑ_2 be two Hilbert spaces and $S\,(\,.\,,.\,)$ a bilinear and bicontinuous form on $\vartheta_1 \times \vartheta_2$ such that there exists a strictly positive constant c satisfying:

H1) $\qquad \forall \ t \in \vartheta_1\,, \ \sup\limits_{y \in \vartheta_2} \dfrac{S\,(t\,,y)}{\|y\|_{\vartheta_2}} \geq c\,\|t\|_{\vartheta_1}\,,$

H2) $\qquad$ *if $y \in \vartheta_2$ is such that $\forall \ t \in \vartheta_1\,, S\,(t\,,y) = 0$, then $y = 0$.*

Then for any linear and continuous form, say $g\,(\,.\,)$ on the space ϑ_2, there exists a unique solution $t \in \vartheta_1$ such that:

$$\forall \ y \in \vartheta_2\,, \ S\,(t\,,y) \;=\; g\,(y)$$

■

In order to apply this lemma to the existence and uniqueness of t^η solution to (II.55), we set:

- $\qquad \vartheta_1 \;=\; H_+^{1/2}(\gamma_2)$

- $\qquad \vartheta_2 \;=\; \vartheta/\vartheta_0$ (quotient space of ϑ by ϑ_0)

- $\qquad \forall \ t \in \vartheta_1\,, \forall \ y = (\mu\,, v_3) \in \vartheta \ \ S\,(t\,,v) \;=\; \left\langle \mu_s + \partial_s v_3\,, t \right\rangle$

- $$\forall\, y = (\mu, v_3) \in \vartheta,\; g\,(y) = -\, k\left(\theta^{\eta}, \mu\right) + \int_{\omega} f_3\, v_3 - \int_{\omega} \left(\mu_{\alpha} + \partial_{\alpha}\, v_3\right) q_{\alpha}^{\eta}$$

where $q_{\alpha}^{\eta} = \dfrac{1}{\eta}\left(\theta_{\alpha}^{\eta} + \partial_{\alpha}\, u_3^{\eta}\right)$.

From the definition of θ^{η} , u_3^{η} and q^{η}, one has the following relation (optimality of (II.52)):

$$\forall\; y \in \vartheta_0,\; g\,(y) = 0$$

Hence $g\,(.)$ is perfectly defined over the quotient space $\vartheta\,/\,\vartheta_0$. It is also true for $S\,(\,.\,,.\,)$, which is such that:

$$\forall\; t \in \vartheta_1,\, \forall\, y \in \vartheta_0 \qquad S\,(t\,,\, y) = 0$$

The linearity and continuity being obvious, let us check the two hypotheses H1 and H2 of Lemma II.1.

i) Checking hypothesis H1. Let us set $y = (\mu\,,\, v_3) \in \vartheta$. Then:

$$\forall\; t \in \vartheta_1,\quad \sup_{y\,\in\,\vartheta/\vartheta_0} \frac{S\,(t\,,\, y)}{\|y\|_{\vartheta/\vartheta_0}} \geq \sup_{y\,\in\,\vartheta} \frac{S\,(t\,,\, y)}{\|y\|_{\vartheta}}$$

because $\|y\|_{\vartheta/\vartheta_0} = \inf_{z\,\in\,\vartheta_0} \|y + z\|_{\vartheta}$.

Therefore:

$$\forall\; t \in \vartheta_1,\quad \sup_{y\,\in\,\vartheta} \frac{\left\langle\mu_s + \partial_s\, v_3\,,\, t\right\rangle}{\|y\|_{\vartheta}} \geq \sup_{\mu\,\in\,W_t} \frac{\left\langle\mu_s\,,\, t\right\rangle}{\|\mu\|_{w_t}}\;.$$

Then for any function – say λ – of the space $H_{00}^{1/2}\,(\gamma_2)$ we can associate a unique element – say μ^{λ} – in the space W_t , such that

$$\begin{cases} \mu_s^{\lambda} = \lambda \quad \text{on}\quad \gamma_2\,, \\[2mm] \left\|\mu^{\lambda}\right\|_{W_t} \leq c\|\lambda\|_{1/2,\,\gamma_2} \end{cases}$$

We can, for instance, use the elasticity operator on ω associated to the bilinear form $k\,(\,.\,,\,.\,)$

defined at (II.37). Then from :

$$\sup_{\lambda \,\in\, H_{00}^{1/2}(\gamma_2)} \frac{\langle \lambda , t \rangle}{\|\lambda\|_{1/2\,,\,\gamma_2}} = \|t\|_{-1/2\,,\,\gamma_2}$$

and noticing that:

$$\sup_{\mu \,\in\, W_t} \frac{\langle \mu_s , t \rangle}{\|\mu\|_{W_t}} \geq \frac{1}{c} \sup_{\lambda \,\in\, H_{00}^{1/2}(\gamma_2)} \frac{\langle \lambda , t \rangle}{\|\lambda\|_{1/2\,,\,\gamma_2}}$$

we deduce:

$$\forall \, t \,\in\, \vartheta_1 \,, \quad \sup_{y \,\in\, \vartheta_2} \frac{S(t,y)}{\|y\|_{\vartheta_2}} \geq c_0 \|t\|_{-1/2\,,\,\gamma_2}$$

with another respect from (if $v_3 \in V_3$ then $v_3|_{\gamma_2} \in H_{00}^{1/2}(\gamma_2)$ and the range of V_3 by this trace is the space $H_{00}^{1/2}(\gamma_2)$):

$$\forall \, t \,\in\, \vartheta_1 \,, \quad \sup_{y \,\in\, \vartheta_2} \frac{S(t,y)}{\|y\|_{\vartheta_2}} \geq \sup_{v_3 \,\in\, V_3} \frac{\langle \partial_s v_3 , t \rangle}{\|v_3\|_{1\,,\,\omega}}$$

$$= \sup_{v_3 \,\in\, V_3} - \frac{\langle \partial_s t , v_3 \rangle}{\|v_3\|_{1\,,\,\omega}} \geq c_1 \|\partial_s t\|_{-1/2\,,\,\gamma_2} \,.$$

Finally:

(II.57) $\quad \forall \, t \,\in\, \vartheta_1 \,, \quad \sup_{y \,\in\, \vartheta_2} \frac{S(t,y)}{\|y\|_{\vartheta_2}} \geq c \left\{ \|t\|_{-1/2\,,\,\gamma_2} + \|\partial_s t\|_{-1/2\,,\,\gamma_2} \right\} \geq c \|t\|_{1/2\,,\,\gamma_2}$

Hence H1 is satisfied.

ii) Checking hypothesis H2. Let us consider an element "y" such that:

$$y = (\mu , v_3) \,\in\, \vartheta_2 \left(= \vartheta / \vartheta_0\right), \ \forall \, t \in \vartheta_1 \,, \ S(t,y) = 0 \,,$$

we deduce that:

$$\forall \, t \,\in\, H_+^{1/2}(\gamma_2) \, \langle \mu_s + \partial_s v_3 , t \rangle = 0$$

or else

$$\mu_s + \partial_s v_3 = 0 \quad \text{on } \gamma_2$$

which implies $y \in \vartheta_0$ or else: $y = 0$ in the quotient space $\vartheta_2 = \vartheta / \vartheta_0$. As a matter of fact, the space ϑ_2 has been defined on purpose.

The two hypotheses H1 and H2 of Lemma II.1 being satisfied, we can claim that there exists a unique solution t^η to the first two equations (II.55) such that $t^\eta \in H_+^{1/2}(\gamma_2)$. The above results are summarized hereafter.

Theorem II.3

Let f_3 be an element of the space $L^2(\omega)$. Then for any $\eta > 0$ there exists a unique solution in the space $W_t \times V_3 \times H_t \times H_+^{1/2}(\gamma_2)$ to the variational equation (II.55) which is also solution to (II.52) (for the components θ^η and u_3^η). ∎

II.3.2.3 Convergence of the modified penalty model

Obviously by convergence we mean that the distance between the solutions to the two models tends to zero when the penalty parameter η tends also to zero. The main result is the following.

Theorem II.4

Let us assume that $f_3 \in L^2(\omega)$ and that $q_\alpha = \partial_\beta m_{\alpha\beta} \in L^2(\omega)$ where $m_{\alpha\beta}$ are the bending moments of Kirchhoff-Love model (see section II.2, formula (II.2)). Then there exists a constant c, independent of η such that:

$$\left\| u_3^\eta - u_3 \right\|_{1,\omega} + \sum_{\alpha=1,2} \left\{ \left\| \theta_\alpha^\eta - \theta_\alpha \right\|_{1,\omega} + \left\| q_\alpha^\eta - q_\alpha \right\|_{-1,\omega} \right\} \leq c \sqrt{\eta}$$

$$\sum_{\alpha=1,2} \left\| q_\alpha^\eta \right\|_{0,\omega} \leq c \quad \text{(uniformly with respect to } \eta\text{)}$$

where $\left(\theta^\eta, u_3^\eta, q^\eta \right)$, (respectively $\left(\theta, u_3, q \right)$) is solution to the modified penalty model (II.55) (respectively the Kirchhoff-Love model (II.36)). Furthermore we set:

$$t = - m_{\alpha\beta} a_\alpha b_\beta \qquad (m_{\alpha\beta} \text{ solution to K.L.})$$

and we assume that $t \in H_+^{1/2}(\gamma_2)$. Then:

$$\| t - t^\eta \|_{1/2,\gamma_2} \leq c\,\eta^{1/4}$$

∎

The proof is rather technical and can be omitted at first reading.

Proof of Theorem II.4

Let us notice that both $\left(\theta^\eta, u_3^\eta\right)$ and $\left(\theta, u_3\right)$ satisfy on γ_2 the constraint of the space ϑ_0. Therefore using the optimality condition of the problem (II.52) leads to (let us recall that for $\alpha = 1,2 : \theta_\alpha + \partial_\alpha u_3 = 0$):

$$K^\eta = k\left(\theta^\eta - \theta, \theta^\eta - \theta\right) + \frac{1}{\eta} \sum_{\alpha=1,2} \left\| \theta_\alpha^\eta - \theta_\alpha + \partial_\alpha \left(u_3^\eta - u_3\right)\right\|_{0,\omega}^2$$

or else:

$$K^\eta = k\left(\theta^\eta - \theta, \theta^\eta - \theta\right) + \frac{1}{\eta} \int_\omega \left(\theta_\alpha^\eta + \partial_\alpha u_3^\eta\right)\left(\theta_\alpha^\eta - \theta_\alpha + \partial_\alpha\left(u_3^\eta - u_3\right)\right)$$

then from the optimality relations of (II.52):

$$K^\eta = -k\left(\theta, \theta^\eta - \theta\right) + \int_\omega f_3 \left(u_3^\eta - u_3\right) \quad .$$

Let us now make the bilinear form $k\,(.\,,.\,)$ explicit:

$$K^\eta = -\int_\omega m_{\alpha\beta}\,\partial_\beta\left(\theta^\eta - \theta\right)_\alpha + \int_\omega f_3\left(u_3^\eta - u_3\right) \quad .$$

But we know that $m_{\alpha\beta} \in L^2(\omega)$ (because $u_3 \in H^2(\omega)$). Furthermore we assumed in the Theorem II.4 that: $\partial_\beta m_{\alpha\beta} \in L^2(\omega)$. Hence, as soon as the boundary of ω is smooth enough, the terms $m_{\alpha\beta}\,b_\beta$ are defined on the boundary as elements of the space $H^{-1/2}(\partial\omega)$. If we notice that:

$$m_{\alpha\beta}\,b_\alpha\,b_\beta = 0 \text{ on } \gamma_1 \cup \gamma_2, \text{ and that } \left(\theta^\eta - \theta\right)_\alpha a_\alpha = \left(\theta^\eta - \theta\right)_s = 0 \text{ on } \gamma_1,$$

we can conclude that:

$$-\int_\omega m_{\alpha\beta}\,\partial_\beta\left(\theta^\eta - \theta\right)_\alpha = -\left\langle m_{\alpha\beta}\,b_\beta\,a_\alpha, \left(\theta^\eta - \theta\right)_s\right\rangle + \int_\omega \partial_\beta m_{\alpha\beta}\left(\theta_\alpha^\eta - \theta_\alpha\right)$$

$(<,>$ denotes the duality between $H_{00}^{1/2}(\gamma_2)$ and $H_{00}^{-1/2}(\gamma_2)$). With another respect from:

$$-\partial_{\alpha\beta}\,m_{\alpha\beta} = f_3 \in L^2(\omega)$$

we obtain the expression:

$$K^\eta = - \left\langle m_{\alpha\beta} \, b_\beta \, a_\alpha \, , \left(\theta^\eta - \theta\right)_s \right\rangle + \int_\omega \partial_\beta \, m_{\alpha\beta} \left(\theta_\alpha^\eta - \theta_\alpha\right) - \int_\omega \partial_{\alpha\beta} \, m_{\alpha\beta} \left(u_3^\eta - u_3\right) \; .$$

Let us now observe that $\partial_{\alpha\beta} \, m_{\alpha\beta} \in L^2(\omega)$ and $\partial_\alpha \, m_{\alpha\beta} \in L^2(\omega)$ (hypotheses included in Theorem II.4) imply that $\partial_\alpha \, m_{\alpha\beta} \, b_\beta$ is defined as an element of the space $H^{-1/2}\left(\partial\omega\right)$. Hence:

$$- \int_\omega \partial_{\alpha\beta} \, m_{\alpha\beta} \left(u_3^\eta - u_3\right) = - \left\langle \partial_\alpha \, m_{\alpha\beta} \, b_\beta \, , u_3^\eta - u_3 \right\rangle + \int_\omega \partial_\alpha \, m_{\alpha\beta} \, \partial_\beta \left(u_3^\eta - u_3\right)$$

and therefore:

$$K^\eta = - \left\langle m_{\alpha\beta} \, b_\beta \, a_\alpha \, , \left(\theta^\eta - \theta\right)_s \right\rangle - \left\langle \partial_\alpha \, m_{\alpha\beta} \, b_\beta \, , u_3^\eta - u_3 \right\rangle + \int_\omega \partial_\alpha \, m_{\alpha\beta} \left[\theta_\beta^\eta - \theta_\beta + \partial_\beta \left(u_3^\eta - u_3\right)\right]$$

But on γ_2 :

$$\left(\theta^\eta - \theta\right)_s = - \partial_s \left(u_3^\eta - u_3\right) \in H_{00}^{1/2}(\gamma_2)$$

which enables one to write (using a distribution derivative on γ_2):

$$K^\eta = - \left\langle \partial_\alpha \, m_{\alpha\beta} \, b_\beta + \partial_s \left(m_{\alpha\beta} \, a_\alpha \, b_\beta\right) , u_3^\eta - u_3 \right\rangle + \int_\omega \partial_\alpha \, m_{\alpha\beta} \left(\theta_\beta^\eta - \theta_\beta + \partial_\beta \left(u_3^\eta - u_3\right)\right)$$

$$= \int_\omega \partial_\alpha \, m_{\alpha\beta} \left[\theta_\beta^\eta - \theta_\beta + \partial_\beta \left(u_3^\eta - u_3\right)\right]$$

because for the Kirchhoff-Love solution the boundary term disappears.

Let us now summarize the previous results. We proved that:

$$\left| K^\eta \right| \le \sum_{\alpha = 1 \, , \, 2} \left\{ \left\| \partial_\beta \, m_{\alpha\beta} \right\|_{0 \, , \, \omega} \; \left\| \theta_\alpha^\eta - \theta_\alpha + \partial_\alpha \left(u_3^\eta - u_3\right) \right\|_{0 \, , \, \omega} \right\}$$

and finally, because of the ellipticity of the bilinear form $k\,(\,.\,,\,.\,)$ on W_t , there exists a strictly positive constant c_0 , independent of η, and such that:

$$\forall \, \xi > 0 \quad \sum_{\alpha = 1, 2} \left[c_0 \left\| \theta_\alpha^\eta - \theta_\alpha \right\|_{1, \omega}^2 + \left(\frac{1}{\eta} - \frac{1}{2 \, \xi} \right) \left\| \theta_\alpha^\eta - \theta_\alpha + \partial_\alpha \left(u_3^\eta - u_3 \right) \right\|_{0, \omega}^2 \right]$$

$$\leq \frac{\xi}{2} \sum_{\alpha = 1, 2} \left\| \partial_\beta \, m_{\alpha\beta} \right\|_{0, \omega}^2 .$$

Choosing for instance $\xi = \eta$, we deduce that:

i)
$$\left\| \theta^\eta - \theta \right\|_{W_t} \leq c_1 \sqrt{\eta}$$

and

ii)
$$\sum_{\alpha = 1, 2} \left\| \theta_\alpha^\eta - \theta_\alpha + \partial_\alpha \left(u_3^\eta - u_3 \right) \right\|_{0, \omega} = \sum_{\alpha = 1, 2} \left\| \theta_\alpha^\eta + \partial_\alpha \, u_3^\eta \right\|_{0, \omega} \leq c_2 \, \eta$$

from which we first derive the following estimate for $\eta \leq 1$:

$$\left\| u_3^\eta - u_3 \right\|_{1, \omega} \leq c_3 \sqrt{\eta}$$

and then, because

$$q_\alpha^\eta = \frac{1}{\eta} \left(\theta_\alpha^\eta + \partial_\alpha \, u_3^\eta \right)$$

we have also:

$$\sum_{\alpha = 1, 2} \left\| q_\alpha^\eta \right\|_{0, \omega} \leq c_4 .$$

Let us now go to the estimate on $q_\alpha^\eta - q_\alpha$. From the definition of these terms, one has:

$$\left| \begin{array}{l} \forall \, \mu \, \in \, \left(H_0^1(\omega) \right)^2 , \quad \int_\omega q_\alpha^\eta \, \mu_\alpha = - k \left(\theta^\eta , \mu \right) , \\[2ex] \forall \, \mu \, \in \, \left(H_0^1(\omega) \right)^2 , \quad \int_\omega q_\alpha \, \mu_\alpha = - k \left(\theta , \mu \right) . \end{array} \right.$$

because we assumed that : $q_\alpha = \partial_\beta \, m_{\alpha\beta} \, \in \, L^2(\omega)$. Hence:

$$\forall \, \mu \, \in \, \left(H_0^1(\omega) \right)^2 , \quad \int_\omega \left(q_\alpha^\eta - q_\alpha \right) \mu_\alpha = - k \left(\theta^\eta - \theta , \mu \right)$$

which leads to the estimate (definition of the norm in $H^{-1}(\omega)$ and continuity of the bilinear form $k(\,.\,,\,.\,)$):

$$\left\| q_\alpha^\eta - q_\alpha \right\|_{-1,\omega} \le c_5 \left\| \theta^\eta - \theta \right\|_{1,\omega} \le c_6 \sqrt{\eta}$$

The last step concerns the estimate on $t^\eta - t$. From the definition of t^η one has:

$$\forall \; Y = (\mu\,,\,v_3) \in \vartheta\,,\; \left\langle \mu_s + \partial_s v_3\,,\, t^\eta \right\rangle = - k\left(\theta^\eta, \mu\right) - \int_\omega q_\alpha^\eta \left(\mu_\alpha + \partial_\alpha v_3\right) + \int_\omega f_3\, v_3$$

But: $t = - m_{\alpha\beta}\, a_\alpha\, b_\beta \in H_+^{1/2}(\gamma_2)$ satisfies also (it is a simple exercise left to the reader):

$$\forall \; Y = (\mu\,,\,v_3) \in \vartheta\,,\; \left\langle \mu_s + \partial_s v_3\,,\, t \right\rangle = - k\left(\theta\,,\, \mu\right) - \int_\omega q_\beta \left(\mu_\beta + \partial_\beta v_3\right) + \int_\omega f_3\, v_3$$

Let us recall that we assumed that $q_\beta \in L^2(\omega)$. From the inequality (II.57) established in the proof of Theorem II.2, we have $(Y = (\mu\,,\,v_3))$:

$$\sup_{Y \in \vartheta} \frac{\left\langle \mu_s + \partial_s v_3\,,\, t^\eta - t \right\rangle}{\|Y\|_\vartheta} \ge c \|t^\eta - t\|_{1/2\,,\,\gamma_2} \quad .$$

But from the above equations characterizing t^η and t we have :

$$\sup_{Y \in \vartheta} \frac{\left\langle \mu_s + \partial_s v_3\,,\, t^\eta - t \right\rangle}{\|Y\|_\vartheta} \le c \left\| \theta^\eta - \theta \right\|_{w_t} + \sup_{Y \in \vartheta} \frac{\int_\omega \left(q_\alpha^\eta - q_\alpha\right)\left(\mu_\alpha + \partial_\alpha v_3\right)}{\|Y\|_\vartheta} \quad .$$

From a basic result due to J.L. Lions [16], for any strictly positive number λ and for any element μ_α of the space $H^1(\omega)$, we can associate two elements – say μ_α^0 and μ_α^1 – such that:

$$\begin{cases} \mu_\alpha = \mu_\alpha^0 + \mu_\alpha^1 \;;\; \mu_\alpha^0 \in H_0^1(\omega)\,,\; \mu_\alpha^1 \in H^1(\omega) \\[2mm] \|\mu_\alpha^0\|_{1,\omega} \le \dfrac{c}{\lambda} \|\mu_\alpha\|_{1,\omega}\,,\; \|\mu_\alpha^1\|_{0,\omega} \le c\,\lambda \|\mu_\alpha\|_{0,\omega} \end{cases}$$

where c does not depend on λ. Hence:

$$\sup_{\mu \in W_t} \frac{\int_\omega \left(q_\alpha^\eta - q_\alpha\right) \mu_\alpha}{\|\mu\|_{W_t}} \leq c \sum_{\alpha = 1, 2} \left\{\frac{1}{\lambda} \|q_\alpha^\eta - q_\alpha\|_{-1, \omega} + \lambda \|q_\alpha^\eta - q_\alpha\|_{0, \omega}\right\}$$

If we choose $\lambda = \eta^{1/4}$, we obtain, assuming here again that $q_\alpha \in L^2(\omega)$) :

(II.58)
$$\sup_{\mu \in W_t} \frac{\int_\omega \left(q_\alpha^\eta - q_\alpha\right) \mu_\alpha}{\|\mu\|_{W_t}} \leq c_7 \, \eta^{1/4} \quad .$$

Let us apply the same result as above to the term $\partial_\alpha v_3$ for any $v_3 \in V_3 \cap H^2(\omega)$, hence $\partial_\alpha v_3 \in H^1(\omega)$. We set:

$$\forall \; v_3 \in V_3 \cap H^2(\omega), \; \partial_\alpha v_3 = v_\alpha^0 + v_\alpha^1 .$$

where:
$$\begin{cases} v_\alpha^0 \in H_0^1(\omega), \; v_\alpha^1 \in H^1(\omega) \\[2mm] \|v_\alpha^0\|_{1, \omega} \leq \frac{c}{\lambda} \|v_3\|_{2, \omega}, \; \|v_\alpha^1\|_{0, \omega} \leq c\lambda \|v_3\|_{1, \omega} \quad . \end{cases}$$

Then using the relation :

$$\partial_\alpha q_\alpha^\eta = \partial_\alpha q_\alpha = - f_3$$

which is derived from (II.42), we obtain from a density argument :

(II.59)
$$\sup_{v_3 \in V_3} \frac{\int_\omega \left(q_\alpha^\eta - q_\alpha\right) \partial_\alpha v_3}{\|v_3\|_{1, \omega}} = \sup_{v_3 \in V_3 \cap H^2(\omega)} \frac{\int_\omega \left(q_\alpha^\eta - q_\alpha\right) \partial_\alpha v_3}{\|v_3\|_{1, \omega}}$$

$$= \sup_{v_3 \in V_3 \cap H^2(\omega)} \frac{\int_\omega \left(q_\alpha^\eta - q_\alpha\right) v_\alpha^1}{\|v_3\|_{1, \omega}} \leq c_7 \lambda$$

Hence from (II.58) and (II.59) with $\lambda = \eta^{1/4}$ we deduce that:

$$\|t^\eta - t\|_{1/2, \gamma_2} \leq c_8 \, \eta^{1/4}$$

Remark II.7

It is certainly possible to improve Theorem II.4 and specially the estimate on $\|t^\eta - t\|$. But this requires additional regularity assumptions on u_3. The method consists in the construction of an asymptotic expansion of $\left(\theta^\eta, u_3^\eta, q^\eta, t^\eta\right)$ with respect to η, including the boundary layer effect. We do not go on in this direction (see Ph. Destuynder [10]) which is very technical and not useful for our purpose in the spirit of this text.   ∎

Remark II.8

When γ_2 is empty, there is no difference between the penalty model (i.e. Reissner-Mindlin-Naghdi model) and the modified version which has been studied in this section. Hence the error estimates of Theorem II.4 apply to this case.   ∎

II.4   Natural duality techniques for the bending plate model

The method described hereafter was originally introduced by Ph. Destuynder and Th. Nevers in [5], [17], [18]. It is directly applied to the Kirchhoff-Love model for bending plate.

Let us come back to the Kirchhoff-Love model described in chapter I and recalled in a functional framework in section II.2. The displacement u_3 is solution to the following set of equations:

$$(\text{II.60}) \quad \begin{cases} u_3 \in V_3^0 = \left\{ v \mid v \in H^2(\omega), \, v = 0 \text{ on } \gamma_0 \cup \gamma_1, \, \partial_\alpha v = 0 \text{ on } \gamma_0 \right\}, \\[4pt] m_{\alpha\beta} = \dfrac{-2 E \varepsilon^3}{3(1 - v^2)} \left\{ (1 - v) \partial_{\alpha\beta} u_3 + v \, \Delta u_3 \, \delta_{\alpha\beta} \right\}, \\[4pt] - \partial_{\alpha\beta} m_{\alpha\beta} = f_3 \text{ on } \omega \\[4pt] m_{\alpha\beta} \, a_\alpha \, b_\beta = 0 \text{ on } \gamma_1 \cup \gamma_2 \\[4pt] \partial_s (m_{\alpha\beta} \, a_\alpha \, b_\beta) + \partial_\alpha m_{\alpha\beta} \, b_\beta = 0 \text{ on } \gamma_2 \end{cases}$$

We already know that (II.60) has a unique solution defined from the variational formulation. Our goal in this section is to construct a new variational formulation which is of the mixed type. The basic point is to introduce the rotation of the unit normal to the medium surface ω as an independent unknown.

Then the constraint:

$$\theta_\alpha + \partial_\alpha u_3 = 0 \qquad \alpha = 1, 2$$

is taken into account by a duality technique. The main difficulty we met is to construct the dual space to the above relation. But first of all let us mention the basic results which will be used in the mathematical analysis. The following Lemma II.2 can be found in R. Temam [13] or V. Girault - P. A. Raviart [14]. Lemma II. 3, is due to L. Tartar but has never been published by its author. Therefore the proof can be found for instance in the book by R. Temam [13] p.20.

Lemma II.2

Let ω be an open set in $\mathbb{R}^2$ with a smooth boundary (for instance piecewise C^1). Then the connected components of the boundary -say $\partial\omega$- are denoted by $\partial\omega_k$, for $k=1$, K. The boundary of ω is assumed to be split into three sub-parts denoted by γ_0, γ_1 and γ_2 (as we did before). Let us introduce a first space by (a $_\alpha$ are the components of the unit tangent to the boundary):

$$X_0 = \left\{ v = (v_\alpha) \, / \, v_\alpha \in H^1(\omega) \; ; \; \text{curl } v = 0 \; ; \; \forall \, k = 1, K \int_{\partial\omega_k} v_\alpha \, a_\alpha = 0 \right\} .$$

It is equipped with the norm of the space $(H^1(\omega))^2$. Let us also define the space :

$$Y_0 = \left\{ v \, / \, v = \text{grad } \phi \, , \, \phi \in H^2(\omega) \right\} .$$

The norm used for Y_0 is the one of the function ϕ in the space $H^2(\omega)$. Then X_0 and Y_0 are isomorphic.

Let us now define a subspace of X_0 by:

$$X_{00} = \left\{ v = (v_\alpha) \, / \, v_\alpha \in H^1_0(\omega) \; ; \; \text{curl } v = 0 \right\} .$$

Then the corresponding subspace of Y_0 (i.e. isomorphic to X_{00}) is:

$$Y_{00} = \left\{ \begin{array}{c} v \, / \, v = \text{grad } \phi \; ; \; \phi \in H^2(\omega) \; ; \; \phi = c_k \text{ on } \partial\omega_k \, , \, \forall \, k = 1, K\text{-}1 \; ; \\ \phi = 0 \text{ on } \partial\omega_K \; ; \; \partial_\alpha\phi \, b_\alpha = 0 \text{ on } \partial\omega \end{array} \right\} .$$

∎

Lemma II.3

The notations of Lemma II.2 are used. Let us consider a vector field say $q=(q_\alpha)$ such that $q = (q_\alpha) \in \left(H^{-1}(\omega)\right)^2$, and satisfying:

(i) $\forall \, v \in X_{00}$, $\langle q_\alpha , v_\alpha \rangle = 0$,

then there exists a unique function - say ψ - such that:

$$q = \text{rot } \psi \quad \text{and} \quad \psi \in L_0^2(\omega) \ .$$

When the boundary of ω is simply connected (i.e. K=1), then condition (i) is equivalent to:

$$\text{div } q = 0 \ ,$$

and the derivatives have to be understood as distributions. When there is more than one connected component on the boundary of ω, equation (i) contains also the condition:

$$\langle q_\alpha b_\alpha , 1 \rangle_k = 0 , \text{ for } k = 1, K\text{-}1,$$

where $\langle , \rangle_k$ denotes the duality between the spaces $H^{1/2}(\partial\omega_k)$ and their dual spaces $H^{-1/2}(\partial\omega_k)$ (the boundary $\partial\omega_k$ is closed, hence the subscript "00" is useless). ■

II.4.1 A mixed variational formulation for Kirchhoff-Love model

It is assumed in this first step, that the boundary of the open set ω is simply connected in order to get rid of an additional difficulty in the application of L. Tartar Lemma (Lemma II.3), that has been recalled previously. The extension to a general case is presented at section II.4.8. The results are the same but with an additional difficulty due to the various components of the boundary.

The shear stress is introduced as an independent unknown by the relation:

(II.61) $\quad q_\alpha = \partial_\beta m_{\alpha\beta} \ .$

It satisfies :

$$\partial_\alpha q_\alpha = \partial_{\alpha\beta} m_{\alpha\beta} = -f_3 \text{ on } \omega \ .$$

Let us now introduce a function φ of the space $H^1(\omega)$ and satisfying the system :

(II.62) $\quad \begin{cases} -\Delta\varphi = f_3 \text{ on } \omega \ , \\ \quad \varphi = 0 \text{ on } \gamma_0 \cup \gamma_1 \ , \\ \quad \dfrac{\partial\varphi}{\partial b} = 0 \text{ on } \gamma_2 \ \left(\dfrac{\partial .}{\partial b} \text{ is the normal derivative along the boundary of } \omega \right) \end{cases}$

If we set:

$$p = q - \text{grad } \varphi \quad \text{with} \quad p = (p_\alpha) \text{ and } q = (q_\alpha) ,$$

we thus obtain:

$$\text{div } p = \partial_\alpha p_\alpha = 0, \text{ and } p_\alpha \in H^{-1}(\omega) .$$

The L. Tartar's Lemma ensures the existence of a function ψ in the space $L_0^2(\omega)$, such that:

$$p = \text{rot } \psi \quad (\text{i.e. } p_1 = -\partial_2 \psi , \ p_2 = \partial_1 \psi).$$

Returning to the definition of q we obtain:

$$(\text{II.63}) \quad \begin{cases} \partial_2 \psi + \partial_\beta m_{\beta 1} = \partial_1 \varphi \text{ on } \omega , \\ -\partial_1 \psi + \partial_\beta m_{\beta 2} = \partial_2 \varphi \text{ on } \omega . \end{cases}$$

furthermore, as on $\gamma_2 : \dfrac{\partial \varphi}{\partial b} = 0$:

$$\partial_\alpha \psi \, a_\alpha = \partial_\beta m_{\beta\alpha} \, b_\alpha = \partial_s \psi \text{ on } \gamma_2$$

where s is the curvilinear abscissa along the boundary of ω. But from the boundary conditions satisfied by the plate solution on γ_2 one has:

$$\partial_s (m_{\beta\alpha} \, a_\alpha \, b_\beta) + \partial_\alpha m_{\beta\alpha} \, b_\beta = 0 \text{ on } \gamma_2$$

Accordingly, we infer:

$$\partial_s (\psi + m_{\beta\alpha} \, a_\alpha \, b_\beta) = 0 \text{ on } \gamma_2 .$$

Finally on each connected component of γ_2, i.e. γ_2^i for i varying from 1 to N, we have:

$$(\text{II.64}) \quad \psi = - m_{\alpha\beta} \, a_\alpha \, b_\beta + c_i \quad i = 1 , N \text{ on } \gamma_2$$

N being the number of connected components of γ_2 on the boundary $\partial\omega$ of ω .

Equations (II.63) lead to the following variational formulation (the edge terms along γ_0 and γ_1 vanish because of the boundary conditions verified by $m_{\alpha\beta}$ and those satisfied by the elements μ of the space W_t (defined below):

$$\forall\,\mu\,\in\,W_t\,,\,-\int_{\omega} m_{\alpha\beta}\,\partial_\alpha\,\mu_\beta\,+\int_{\gamma_2} m_{\alpha\beta}\,b_\beta\,\mu_\alpha\,+\int_{\gamma_2}\psi\,\text{curl}\,\mu\,+\int_{\gamma_2}\psi\,\mu_s\,=\,\int_{\omega}\partial_\alpha\,\varphi\,\mu_\alpha\,,$$

where:

$$\overset{\text{definition}}{\text{curl}\,\mu\;\equiv\;\partial_1\,\mu_2\,-\,\partial_2\,\mu_1}\;,$$

and the space W_t is given by:

$$W_t\;=\;\left\{\mu\,|\,\mu\,=\,(\mu_\alpha)\,\in\,\left(H^1\,(\omega)\right)^2\;;\;\mu_\alpha=0\;\text{on}\;\gamma_0\;;\;\mu_\alpha\,a_\alpha=0\;\text{on}\;\gamma_1\right\}\,.$$

Finally the notation μ_s denotes $\mu_\alpha a_\alpha$ which is the tangential component of μ along the boundary of ω. Using (II.64) yields:

$$\forall\,\mu\,\in\,W_t\,,\,\int_{\omega} m_{\alpha\beta}\,\partial_\alpha\,\mu_\beta\,-\int_{\omega}\psi\,\text{curl}\,\mu\,-\,\sum_{i=1,N} c_i\int_{\gamma_2^i}\mu_s\,=\,-\int_{\omega}\partial_\alpha\,\varphi\,\mu_\alpha\;.$$

Furthermore, the solution to Kirchhoff-Love model satisfies:

$$\int_{\gamma_2^i}\theta_s\,=\,-\int_{\gamma_2^i}\partial_s\,u_3\,=\,0\;\;\forall\;i=1,N$$

because each connected component of γ_2 is either closed (see section II.4.8), or surrounded by portions of boundary of γ_0 or γ_1 on which $u_3=0$. Furthermore, the solution to the Kirchhoff-Love model verifies:

$$\text{curl}\,\theta\,=\,\partial_1\,\theta_2\,-\,\partial_2\,\theta_1\,=\,0\;.$$

Finally the element $(\theta\,,\,\psi\,,\,c)$ of the space $W_t\,\times\,L^2\,(\omega)\,\times\,\mathbb{R}^N$ is a solution to the following system:

$$\begin{cases} \forall \; \mu \; \in \; W_t \,, \; k\left(\theta\,,\mu\right) - \int_\omega \psi \; \text{curl} \; \mu \; - \sum_{i=1,N} c_i \int_{\gamma_2^i} \mu_s \; = \; - \int_\omega \partial_\alpha \, \varphi \, \mu_\alpha \\ \\ \forall \; q \; \in \; L^2\left(\omega\right), \int_\omega q \; \text{curl} \; \theta \; = \; 0 \,, \\ \\ \forall \; M \; \in \; \mathbb{R}^N \,, \; M_i \int_{\gamma_2^i} \theta_s \; = \; 0 \end{cases}$$

(II.65)

where the bilinear form k (. , .) is the same as the one used in the previous section i.e.:

$$k\left(\theta\,,\mu\right) \; = \; \int_\omega m_{\alpha\beta} \, \partial_\alpha \, \mu_\beta \; = \; \int_\omega \frac{2 \, E \, \epsilon^3}{3\left(1 - \nu^2\right)} \left\{ \left(1 - \nu\right) \gamma_{\alpha\beta}\left(\theta\right) \gamma_{\alpha\beta}\left(\mu\right) \; + \; \nu \, \gamma_{\lambda\lambda}\left(\theta\right) \gamma_{\nu\nu}\left(\mu\right) \right\} .$$

This is a conventional mixed formulation. The existence and uniqueness of a solution and the interpretation of this solution are given in section II.4.2 (as a matter of fact there is no uniqueness for the constants C). Let us first make some comments.

Remark II.8
It is pointed out that the equation :

$$\int_{\gamma_2^i} \theta_s \; = \; 0 \qquad i = 1 \,, N$$

is not local. This means that the degrees of freedom (in a finite element approximation) of θ_s along γ_2^i are related. Accordingly, the matrix associated with linear system (II.65) has an impracticable structure (very large bandwidth on a free edge). This difficulty can be overcome by a duality – like solving algorithm (minimization of the dual functional with respect to ψ and c, in the space $L_0^2\left(\omega\right) \times \mathbb{R}^N$). Another process consists in eliminating the dual variables (φ and c) by regularizing the dual model (see for details on such a method M. Bercovier [19]). Then we return to the Mindlin model recalled in section II.3 as we see in section II.4.7. ∎

Remark II.9
Once ψ , φ , θ and c have been computed, u_3 can be characterized by the relation:

$$\theta_\alpha \; = \; - \partial_\alpha \, u_3 \qquad \text{on} \;\; \omega$$

which implies, and it is only an implication :

(II.66)

$$
\begin{cases}
- \Delta\, u_3 = \operatorname{div} \theta & \text{on } \omega, \\[2mm]
u_3 = 0 & \text{on } \gamma_0 \cup \gamma_1, \\[2mm]
\dfrac{\partial\, u_3}{\partial\, b} = \theta_\alpha\, b_\alpha & \text{on } \gamma_2.
\end{cases}
$$

Equations (II.66) of course have a unique solution u_3 in the space $H^1(\omega)$ (classical) and, which is remarkable, the model to be solved is the same as that characterizing φ (except the right-hand-side). It is therefore not necessary to refactorize the associated matrix. The reader should notice that the construction of the mixed model suggested in the present section is quite formal (although if we look carefully at the different steps, they can be completely justified). The mathematical proof of the validity of our model is contained in the next section. ∎

II.4.2   Existence and uniqueness of solution to the mixed formulation

Let us introduce a few notations which are helpful in the following. First of all we recall the definition of the bilinear form $k(.,.)$:

$$\forall\ \theta, \mu \in W_t : k(\theta, \mu) = \int_\omega m_{\alpha\beta}\, \partial_\alpha\, \mu_\beta$$

$$= \int_\omega \frac{2\,E\,\epsilon^3}{3(1-v^2)}\left\{(1-v)\,\gamma_{\alpha\beta}(\theta)\,\gamma_{\alpha\beta}(\mu)\ +\ v\,\gamma_{\lambda\lambda}(\theta)\,\gamma_{vv}(\mu)\right\}\ .$$

Let us set:

$$M = L^2(\omega) \times \mathbb{R}^N\ ,$$

and for any element $\Lambda = (\psi, c) \in M$ and $\mu \in W_t$, we define the bilinear form:

$$b(\mu, \Lambda) = -\int_\omega \psi\, \mathrm{curl}\, \mu\ -\ \sum_{i=1,N} c_i \int_{\gamma_2} \mu_s$$

Finally, φ being the unique solution to system (II.62), $\varphi \in H^1(\omega)$, we set:

$$\forall\ \mu \in W_t,\ g(\mu) = -\int_\omega \partial_\alpha\, \varphi\, \mu_\alpha$$

We can now formulate the main result of this section.

Theorem II.5
Let us set the mixed formulation:

$$\begin{cases} \text{find } (\theta, \Lambda) \in W_t \times M \quad \text{such that :} \\ \forall\ \mu \in W_t,\ k(\theta, \mu) + b(\mu, \Lambda) = g(\mu), \\ \forall\ \Xi \in M\ ,\ b(\theta, \Xi) = 0. \end{cases}$$

Then there exists a solution (θ, Λ) to this model. The component θ is unique and $\Lambda = (\psi, c)$ is uniquely defined up to a constant for ψ .

If $N > 0$, this constant can be eliminated by choosing one of the N constants c_i equal to zero. When $N = 0$, we can look for ψ in the space $L^2_0(\omega)$ and the space M is replaced by $L^2_0(\omega)$.   ■

The proof is obtained from Brezzi Theorem [20] which we recall hereafter in a simpler formulation, but sufficient for our purpose.

Theorem II.6 (Simplified version of Brezzi Theorem)

Let Σ , ϑ be two Hilbert spaces and a (. , .), respectively b (. , .), a bilinear form on $\vartheta \times \vartheta$, respectively on $\Sigma \times \vartheta$. They are supposed to satisfy the following properties:

(ellipticity) $\qquad\qquad\qquad \forall \ v \in \vartheta , a\left(v , v\right) \geq c_0 \ \|v\|_\vartheta^2$

(continuity) $\qquad\qquad\qquad \forall \ u , v \in \vartheta , |a(u , v)| \leq c_1 \ \|u\| \ \|v\|_\vartheta$

(inf-sup condition) $\qquad\qquad \forall \ p \in \Sigma , \ \sup_{v \in \vartheta} \dfrac{b\left(p , v\right)}{\|v\|_\vartheta} \geq c_2 \ \|p\|_\Sigma$

(continuity) $\qquad\qquad\qquad \forall \ p , v \in \Sigma \times \vartheta, \ \ |b\left(p , v\right)| \leq c_3 \ \|p\|_\Sigma \|v\|_\vartheta$

Then for any linear and continuous form g, respectively h, on ϑ, respectively Σ, there exists a unique element (u, p) in the space $\vartheta \times \Sigma$ such that:

$$\begin{cases} \forall \ v \in \vartheta, a(u , v) + b(p , v) = g(v) \\ \forall \ q \in \Sigma, b(q , u) = h(q) \end{cases}$$

∎

Proof of Theorem II.5.
We set:

$$\Sigma = M = L_0^2(\omega) \times \mathbb{R}^N$$

where

$$L_0^2(\omega) = \left\{ f \in L^2(\omega) , \ \int_\omega f = 0 \right\} ,$$

and the space ϑ is chosen equal to W_t which is:

$$W_t = \left\{ \mu \,|\, \mu = (\mu_\alpha) \in \left(H^1(\omega)\right)^2 , \ \mu_\alpha = 0 \text{ on } \gamma_0 , \ \mu_s = \mu_\alpha a_\alpha = 0 \text{ on } \gamma_1 \right\} .$$

Furthermore the bilinear forms a (. , .) and b (, ,) of Theorem II.6 are respectively k (,) and b (,) of Theorem II.5. Finally we have h = 0 and g (.) being the linear form from

Theorem II.5. The continuities are easy to check. The ellipticity of k (,) is a direct consequence of Korn inequality in plane elasticity. The only difficulty is then to prove the "sup" condition on b (,). First of all, let us notice that $\left(H_0^1(\omega)\right)^2$ is included in W_t. Hence for any $\Lambda \in M$:

$$\sup_{\mu \in W_t} \frac{b(\mu, \Lambda)}{\|\mu\|_{W_t}} \geq \sup_{\mu \in (H_0^1(\omega))^2} \frac{b(\mu, \Lambda)}{\|\mu\|_{W_t}}$$

But :

$$\forall \mu \in \left(H_0^1(\omega)\right)^2, \ b(\mu, \Lambda) = \int_\omega \psi \ \text{curl} \ \mu = - \int_\omega \text{rot} \ \psi \ \mu$$

and thus:

$$\sup_{\mu \in W_t} \frac{b(\mu, \Lambda)}{\|\mu\|_{W_t}} \geq c \sum_{\alpha = 1, 2} \|\partial_\alpha \psi\|_{-1, \omega}$$

where $\Lambda = (\psi, c) \in L_0^2(\omega) \times \mathbf{R}^N$. Because the norm (cf. Temam for instance [13]) :

$$\psi \in L_0^2(\omega) \rightarrow \sum_{\alpha = 1, 2} \|\partial_\alpha \psi\|_{-1, \omega}$$

is equivalent, on $L_0^2(\omega)$, to the one induced by $L^2(\omega)$, we deduce :

(II.67)
$$\sup_{\mu \in W_t} \frac{b(\mu, \Lambda)}{\|\mu\|_{W_t}} \geq c \ \|\psi\|_{0, \omega} \ .$$

For each connected component of the boundary γ_2, say γ_2^i i = 1 , N, we consider a smooth function, denoted g_i, such that:

$$\int_{\gamma_2^i} g_i = 1$$

and g_i vanishes at both ends of γ_2^i (if they exist). Then we define $\mu_i \in W_t$ such that :

$$\begin{cases} \mu_s^i = c^i \, g_i & \text{on } \gamma_2^i \quad \left(\mu_s^i = \mu_\alpha \, a_\alpha\right) \\ \mu_n^i = 0 & \text{on } \gamma_2^i \quad \left(\mu_n^i = \mu_\alpha \, b_\alpha\right) \\ \mu^i = 0 & \text{on } \gamma_2^j \quad \forall j = 1, N, \neq i \end{cases}$$

and:

$$\|\mu^i\|_{W_t} \leq c \|c^i \, g_i\|_{1/2, \gamma_2^i} = c \, |c^i| \, \|g_i\|_{1/2, \gamma_2^i} \quad .$$

Finally, denoting by : $\Lambda = (\psi, c) \in L_0^2(\omega) \times \mathbb{R}^N$, we obtain :

(II.68) $$\sup_{\mu \in W_t} \frac{b(\mu, \Lambda)}{\|\mu\|_{W_t}} \geq \frac{b(\mu^i, \Lambda)}{\|\mu^i\|_{W_t}} \geq c_0 \, |c^i| - c_1 \, \|\psi\|_{0, \omega}$$

where c_0 and c_1 are two positive constants. Then (II.67) and (II.68) lead to:

$$\sup_{\mu \in W_t} \frac{b(\mu, \Lambda)}{\|\mu\|_{W_t}} \geq c_2 \, \|\Lambda\|_M$$

which permits one to apply Theorem II.6 in order to prove Theorem II.5.

Finally, let us point out that we chose: $M = L_0^2(\omega) \times \mathbb{R}^N$ in order to ensure the uniqueness of (θ, Λ). ■

II.4.3 Computation of the deflection u_3

The rotations θ_α are computed from the formulation given in Theorem II.5. Then the relation :

$$\theta_\alpha = -\partial_\alpha u_3 \qquad \alpha = 1, 2$$

implies that u_3 is solution of:

(II.69) $$\begin{cases} u_3 \in V_3, \text{ such that :} \\ \forall \ v_3 \in V_3, \ \int_\omega \partial_\alpha u_3 \, \partial_\alpha v_3 = - \int_\omega \theta_\alpha \, \partial_\alpha v_3 \end{cases}$$

This is a classical variational equation and the existence and uniqueness of a solution is obtained by applying Lax-Milgram Theorem. One could object that (II.69) is only a consequence of the relation $\theta_\alpha = - \partial_\alpha u_3$ but it is not equivalent. This remark is correct. But one has to remember that we know that u_3 exists as the solution to Kirchhoff-Love model. Thus (II.69) is just a method for computing it.

II.4.4   *How to be sure we solved the right model (interpretation of the model)*

Systems (II.62) for φ , (II.65) for $\left(\theta , \psi , c\right)$ and (II.69) for u_3 , are satisfied by the solution u_3 to Kirchhoff-Love model if we set :

(II.70)

$$\left|\begin{array}{l} \theta_\alpha = - \partial_\alpha u_3 \quad \left(\in W_t\right), \\[2ex] \varphi \text{ solution to (II.62) ,} \\[2ex] \psi \text{ being such that } \operatorname{rot} \psi = \left(\partial_\beta m_{\alpha\beta}\right) - \operatorname{grad} \varphi \\[2ex] \text{where} \\[2ex] m_{\alpha\beta} = \dfrac{-2\,E\,\varepsilon^3}{3\left(1 - v^2\right)} \left\{(1 - v)\, \partial_{\alpha\beta}\, u_3 + v\, \Delta\, u_3\, \delta_{\alpha\beta}\right\}, \\[2ex] \text{and} \\[2ex] c_i = \left(\psi + m_{\alpha\beta}\, a_\alpha\, b_\beta\right) \quad \text{on} \quad \gamma_2^i \quad i = 1\,,N \quad. \end{array}\right.$$

By eliminating the function ψ from the last relation (taking the derivative with respect to the curvilinear abscissa s) and the relation :

$$(\operatorname{rot} \psi)_\alpha\, b_\alpha = \frac{\partial \psi}{\partial s} \quad,$$

we deduce the Kirchhoff-Love boundary condition on a free edge:

$$\partial_s\!\left(m_{\alpha\beta}\, a_\alpha b_\alpha\right) + \partial_\alpha\, m_{\alpha\beta}\, b_\beta = 0 .$$

As we proved that the solution to (II.62), (II.65) and (II.69) exists and is unique, it is the one derived from u_3 solution of Kirchhoff-Love through the formulae (II.70).

Remark II.10
There are several possibilities for deriving a mixed formulation from Kirchhoff-Love model. Basically they can be obtained by different choices of the boundary condition for the potential function φ.

For instance for arbitrarily connected boundary of ω, we choose (see section II.4.8) φ equal to a constant on each connected component of $\partial\omega \cap (\gamma_0 \cup \gamma_1)$ instead of $\varphi = 0$ in the present case. The advantage of choosing $\varphi = 0$ on $\gamma_0 \cup \gamma_1$ is to lead to an identical operator for φ and u_3 .   ∎

II.4.5   *What is the meaning of ψ and when is it zero ?*

The function ψ that appears in system (II.65) is a Lagrange multiplier of the relation:

$$\text{curl } \theta = 0 \ .$$

As a matter of fact, one has:

$$(\text{rot } \psi)_\alpha = \partial_\beta m_{\alpha\beta} - \partial_\alpha \varphi$$

or else, by substituting $m_{\alpha\beta}$ versus u_3 :

$$(\text{rot } \psi)_\alpha = - \frac{2\,E\,\varepsilon^3}{3\,(1 - v^2)} \, \partial_\alpha \Delta u_3 - \partial_\alpha \varphi$$

and setting :

$$k = - \frac{2\,E\,\varepsilon^3}{3\,(1 - v^2)} \, \Delta u_3 - \varphi$$

one has :

$$\text{rot } \psi = \text{grad } k \ .$$

Let us now observe that:

$$\Delta k = - \frac{2\,E\,\varepsilon^3}{3\,(1 - v^2)} \, \Delta^2 u_3 - \Delta \varphi = - f + f = 0 \ .$$

Furthermore :

$$k = - \frac{2\,E\,\varepsilon^3}{3\,(1 - v^2)} \, \Delta u_3 \quad \text{on } \gamma_0 \cup \gamma_1$$

$$\frac{\partial k}{\partial b} = \partial_\alpha k \, b_\alpha = \frac{2\,E\,\varepsilon^3}{3\,(1 - v^2)} \, \frac{\partial \Delta u_3}{\partial b} \quad \text{on } \gamma_2$$

But $\Delta\,u_3$ is the mean curvature (see chapter VI) of the deformed medium surface of the plate. Hence $k = 0$ on $\gamma_0 \cup \gamma_1$ if the deformed plate in the vicinity of this boundary is planar (because one of the curvatures along $\gamma_0 \cup \gamma_1$ remains zero). The term $\dfrac{\partial k}{\partial b}$ is proportional to the transverse shear stress along the boundary γ_2 $\left(Q_\alpha = -\dfrac{2\,E\,\varepsilon^3}{3\,(1 - v^2)}\,\partial_\alpha\,\Delta\,u_3 \right)$. Hence, the relation $\dfrac{\partial k}{\partial b} = 0$ on γ_2, is true if $Q_\alpha\,b_\alpha = 0$ on this boundary. As a conclusion we can say that: $k = 0$ and therefore $\psi = 0$ if the plate remains planar in the vicinity of the boundary $\gamma_0 \cup \gamma_1$ and if $Q_\alpha\,b_\alpha = 0$ on γ_2 (no transverse loading near the free edge γ_2). Obviously these conclusions are limited to homogeneous and isotropic materials. For composites the problem is different (see for instance chapter V). Finally we can say that ψ is very much connected to the warping of the plate near the boundary.

II.4.6 Non-homogeneous boundary conditions

There are no special difficulties when the boundary conditions are non-homogeneous. Let us for instance consider the following case:

•   on $\gamma_1 \cup \gamma_2$ $m_{\alpha\beta}\,b_\alpha\,b_\beta\ =\ C$

which is a torque distributed along $\gamma_1 \cup \gamma_2$.

•   on γ_2 $-\left[\partial_s\left(m_{\alpha\beta}\,a_\alpha\,b_\beta\right) + \partial_\beta\,m_{\alpha\beta}\,b_\beta\right]\ =\ T$

which is a torque normal to the boundary γ_2.

The Kirchhoff-Love model is well posed, as soon as C and T are smooth enough functions (the regularity $L^2\left(\partial\omega\right)$ is not necessary but sufficient). A simple exercise left to the reader shows that the first equation (II.65) is thus modified as follows in order to take into account the loads:

$$\forall\ \mu \in W_t\ \ k\left(\theta\,,\mu\right) - \int_\omega \phi\,\mathrm{curl}\,\mu - \sum_{i\,=\,1,\,N} c_i \int_{\gamma_2} \mu_s + \int_\omega \mathrm{grad}\,\phi\,\mu\ =\ \int_{\gamma_1\cup\gamma_2} C\left(\mu_\alpha\,b_\alpha\right)$$

$$\forall\ v \in V_3,\ \int_\omega \mathrm{grad}\,\phi\,\mathrm{grad}\,v\ =\ \int_{\gamma_2} T\,v + \int_\omega f\,v$$

where we recall that $\mu_s = \mu_\alpha a_\alpha$.

II.4.7 *The revisited modified Reissner-Mindlin-Naghdi model*

Let us come back to the penalty model that we introduced in section II.3.2, with the modification on the free edge γ_2 . Let us recall the variational formulation (II.55) (the space $H_+^{1/2}(\gamma_2)$ has been defined previously at section III.3.2):.

$$
(\text{II.71}) \quad
\begin{cases}
\text{find } \left(\theta^\eta, u_3^\eta, q^\eta, t^\eta\right) \in W_t \times V_3 \times H_t \times H_+^{1/2}(\gamma_2) : \\[2mm]
\forall \ \mu \in W_t, \ k\left(\theta^\eta, \mu\right) + \int_\omega q_\alpha^\eta \mu_\alpha + \left\langle t^\eta, \mu_s\right\rangle = 0 , \\[2mm]
\forall \ v_3 \in \vartheta_3, \ \int_\omega q_\alpha^\eta \partial_\alpha v_3 + \left\langle t^\eta, \partial_s v_3\right\rangle = \int_\omega f_3 \, v_3 , \\[2mm]
\forall \ p \in H_t, \ \eta \int_\omega q_\alpha^\eta p_\alpha = \int_\omega p_\alpha \left(\theta_\alpha^\eta + \partial_\alpha u_3^\eta\right) , \\[2mm]
\forall \ r \in H_+^{1/2}(\gamma_2), \ \left\langle r, \theta_s^\eta + \partial_s u_3^\eta\right\rangle = 0 ,
\end{cases}
$$

where $H_t = \left[L^2(\omega)\right]^2$, W_t and V_3 being defined at (II.39). Existence, uniqueness and convergence (when $\eta \longrightarrow 0$) of a solution have been extensively established in section (II.3.2). A new formulation, equivalent to (II.71), is derived hereafter. It is more adapted to numerical approximations as we show in Chapter III. As a matter of fact, it is a natural extension of the **natural duality technique** introduced in section II.3. It appears that the modification on the free boundary γ_2 is quite natural and is very much linked to the value of the function ψ introduced in the natural duality formulation. Let us start with a technical result (cf. R. Temam [13]).

Lemma II.4
We assume that the boundary of ω is simply connected. Let $q \in H_t = \left[L^2(\omega)\right]^2$. Then there exist two functions φ and ψ in the space $H^1(\omega)$ such that:

$$\begin{cases} q = \text{grad } \varphi + \text{rot } \psi \\[2ex] \varphi = 0 \quad \text{on } \gamma_0 \cup \gamma_1 \,, \ \dfrac{\partial \varphi}{\partial b} = \partial_\alpha \varphi \, b_\alpha = 0 \ \text{ on } \gamma_2 \\[2ex] \Delta \varphi = \text{div } q \quad \text{(being given !)} \end{cases}$$

Furthermore ψ *is unique if we add the condition* $\psi \in L_0^2(\omega)$. *The function* φ *is always unique.*   ∎

Sketch of the proof (which is quite obvious)

First of all, q being given, there exists a unique function φ such that :

$$\begin{cases} \varphi \in H^1(\omega) \\[2ex] \Delta \varphi = \text{div } q \\[2ex] \varphi = 0 \quad \text{on } \gamma_0 \cup \gamma_1 \,, \ \dfrac{\partial \varphi}{\partial b} = 0 \ \text{ on } \gamma_2 \end{cases}$$

Then the vector :

$$p = q - \text{grad } \varphi \ ,$$

is such that :

$$\text{div } p = 0 \ .$$

Hence, there exists a function ψ (see Lemma II.3) defined up to a constant and such that :

$$p = \text{rot } \psi \ .$$

Finally :

$$q = \text{grad } \varphi + \text{rot } \psi$$

and ψ is unique if we prescribe the condition :

$$\int_\omega \psi = 0 \ .$$

Let us now (because of Lemma II.3) set :

$$q^\eta \ = \ \text{grad} \ \varphi^\eta + \text{rot} \ \psi^\eta \quad \text{on} \ \omega \ .$$

Then system (II.71) is equivalent to :

$$\text{find} \ \left(\theta^\eta \ , u_3^\eta \ , q^\eta \ , t^\eta\right) \ \in \ W_t \times V_3 \times V_3 \times H^1 \left(\omega\right)^0 \times H_+^{1/2} \left(\gamma_2\right) \ \text{such that :}$$

(II.72)
$$\begin{cases}
\forall \ \mu \ \in \ W_t \ , k\left(\theta^\eta \ , \mu\right) + \int_\omega \text{grad} \ \varphi^\eta \cdot \mu \ + \int_\omega \text{rot} \ \psi^\eta \cdot \mu + \left\langle t^\eta \ , \mu_s\right\rangle \ = \ 0 \\[2mm]
\forall \ v_3 \ \in \ V_3 \ , \int_\omega \text{rot} \ \psi^\eta \cdot \text{grad} \ v_3 + \int_\omega \text{grad} \ \varphi^\eta \cdot \text{grad} \ v_3 + \left\langle t^\eta, \partial_s \ v_3\right\rangle = \int_\omega f_3 \ v_3, \\[2mm]
\forall \ p \ \in \ H^1 \left(\omega\right)^0 \ , \eta \int_\omega \left[\text{grad} \ \varphi^\eta + \text{rot} \ \psi^\eta\right] \text{rot} \ p \ = \ \int_\omega \text{rot} \ p \left[\theta^\eta + \text{grad} \ u_3^\eta\right] , \\[2mm]
\forall \ \lambda \ \in \ V_3 \ , \eta \int_\omega \left[\text{grad} \ \varphi^\eta + \text{rot} \ \phi^\eta\right] \text{grad} \ \lambda \ = \ \int_\omega \text{grad} \ \lambda \left[\theta^\eta + \text{grad} \ u_3^\eta\right] , \\[2mm]
\forall \ r \ \in \ H_+^{1/2} \left(\gamma_2\right), \left\langle r \ , \theta_s^\eta + \partial_s \ u_3^\eta\right\rangle = 0 \ .
\end{cases}$$

where we set : $H^1 \left(\omega\right)^0 \ = \ H^1 \left(\omega\right) \cap L_0^2 \left(\omega\right)$. Let us now notice that:

$$\forall \ \mu \ \in \ W_t \ , \int_\omega \text{rot} \ \psi^\eta \cdot \mu \ = \ - \int_{\gamma_2} \psi^\eta \ \mu_s \ - \int_\omega \psi^\eta \ \text{curl} \ \mu \ ,$$

$$\forall \ v_3 \ \in \ V_3 \ , \int_\omega \text{rot} \ \psi^\eta \cdot \text{grad} \ v_3 \ = \ - \int_{\gamma_2} \psi^\eta \ \partial_s \ v_3 \ .$$

Hence if we set :

$$z^\eta \ = \ t^\eta \ - \ \psi^\eta \ \in \ H_+^{1/2} \left(\gamma_2\right) \ ,$$

system (II.72) is also equivalent to:

find $\left(\theta^\eta, u_3^\eta, \varphi^\eta, \psi^\eta, z^\eta\right) \in W_t \times V_3 \times V_3 \times H^1(\omega)^0 \times H_+^{1/2}(\gamma_2)$ such that :

(II.73)

$$\begin{cases} \forall \mu \in W_t,\ k\left(\theta^\eta, \mu\right) - \int_\omega \psi^\eta\, \text{curl}\, \mu + \int_\omega \text{grad}\, \varphi^\eta \cdot \mu + \langle z^\eta, \mu_s\rangle = 0, \\[2mm] \forall\ v_3 \in V_3,\ \int_\omega \text{grad}\, \varphi^\eta \cdot \text{grad}\, v_3 + \langle z^\eta, \partial_s v_3\rangle = \int_\omega f_3\, v_3, \\[2mm] \forall\ q \in H^1(\omega)^0,\ \eta \int_\omega \left[\text{grad}\, \varphi^\eta + \text{rot}\, \psi^\eta\right] \text{rot}\, q = \int_\omega \text{rot}\, q\left[\theta^\eta + \text{grad}\, u_3^\eta\right], \\[2mm] \forall\ \lambda \in V_3,\ \eta \int_\omega \left[\text{grad}\, \varphi^\eta + \text{rot}\, \psi^\eta\right] \text{grad}\, \lambda = \int_\omega \text{grad}\, \lambda\left[\theta^\eta + \text{grad}\, u_3^\eta\right], \\[2mm] \forall\ r \in H_+^{1/2}(\gamma_2),\ \langle r, \theta_s^\eta + \partial_s u_3^\eta\rangle = 0. \end{cases}$$

Remark II.11

System (II.73) would have been exactly the same if we had started from Reissner-Mindlin-Naghdi model without modifications. But the meaning of the variable z^η would be different. Unfortunately (II.73) is a little bit complicated and there is no clear advantage compared to the mixed formulation (II.62) - (II.65) - (II.69). Hence (II.73) is forgotten in the following. Nevertheless, let us point out that in a clamped case a lot of things are simpler, due to the orthogonality of rot ψ and grad φ in $L^2(\omega)$ and because there is no free edge $\gamma_2 = \varnothing$. For instance (II.73) leads to three decoupled problems:

(II.74)

$$\begin{cases} \text{find } \varphi^\eta \in V_3 \text{ such that :} \\[2mm] \forall\ v_3 \in V_3,\ \int_\omega \text{grad}\, \varphi^\eta \cdot \text{grad}\, v_3 = \int_\omega f_3\, v_3. \end{cases}$$

(II.75)

$$\begin{cases} \text{Then : find } \left(\theta^\eta, \psi^\eta\right) \in W_t \times H^1(\omega)^0 \text{ such that :} \\[2mm] \forall\ \mu \in W_t,\ k\left(\theta^\eta, \mu\right) - \int_\omega \psi^\eta\, \text{curl}\, \mu = -\int_\omega \text{grad}\, \varphi^\eta \cdot \mu, \\[2mm] \forall\ q \in H^1(\omega)^0,\ \eta \int_\omega \text{rot}\, \psi^\eta \cdot \text{rot}\, q = \int_\omega q\, \text{curl}\, \theta^\eta. \end{cases}$$

(II.76)

$$\text{And finally : find } u_3^\eta \in V_3 \text{ such that :}$$

$$\forall \; \lambda \in V_3 \,, \int_\omega \text{grad } u_3^\eta \cdot \text{grad } \lambda \;\; = \;\; \int_\omega \left[\eta \text{ grad } \varphi^\eta - \theta^\eta \right] \cdot \text{grad } \lambda$$

where $V_3 = H_0^1(\omega)$, $W_t = \left(H_0^1(\omega) \right)^2$. In this particular case the model obtained is similar to the one suggested independently by F. Brezzi and M. Fortin [21]. But to our opinion this is not the right formulation to be considered. ■

II.4.8 Extension to a multi-connected boundary

First of all it is necessary to point out that φ is no more autonomous. Or at least, not completely. This is certainly the main difference between this section and the previous one. Let us point out again that up to now we assumed that the boundary of ω was simply connected in order to apply the simplest version of Tartar's lemma (see lemma II.3). In this section we get rid of this restriction. More precisely, we discuss how to extend the obtained results to a multi-connected boundary of ω.

II.4.8.1 Characterization of φ, ψ and θ

First of all, let us make explicit the notations. We denote by $\delta\omega_k$ the K connected components of the boundary $\delta\omega$ of ω. Then we introduce on each of these components, the sub-parts of γ_0 and γ_1, say γ_{0k} and γ_{1k}, on which the plate is respectively clamped and simply supported. Let us now define K-1 functions denoted by g_k, for $k = 1$, K-1. Implicitly, we can say that $g_K = 0$. Functions g_k are such that :

(II.77)

$$-\Delta g_k = 0 \text{ on } \omega,$$

$$g_k = 1 \text{ on } \gamma_{0k} \text{ and } \gamma_{1k} \, ; \; g_k = 0 \text{ on } \gamma_{0j} \cup \gamma_{1j} \, , \, j \neq k \\ j = 1, K-1$$

$$\frac{\partial g}{\partial b} = 0 \text{ on } \gamma_2 \text{ (the free edge) where as usual } \frac{\partial \cdot}{\partial b} \text{, is the normal derivative .}$$

The system II.77 defines g_k as a unique function of the space $H^1(\omega)$. Obviously, if $K = 1$, there is no need for the g_k functions. Let us now introduce the transverse shear stress by:

$$q_\alpha = \partial_\beta\, m_{\alpha\beta}\ ,\qquad \alpha = 1\,,\, 2,$$

where the bending moments are given by:

$$m_{\alpha\beta} = \frac{-2\,E\,\varepsilon^3}{3\left(1 - v^2\right)}\left\{(1 - v)\,\partial_{\alpha\beta}\, u_3 + v\,\Delta\, u_3\,\delta_{\alpha\beta}\right\},$$

u_3 being the solution to the Kirchhoff-Love model. It is worth noticing that, because u_3 is in the space $H^2(\omega)$, and the given function f_3 is in the space $L^2(\omega)$, one has:

$$q_\alpha \in H^{-1}(\omega)\,,\ \operatorname{div} q = \partial_\alpha\, q_\alpha = -f_3 \in L^2(\omega)\ .$$

Therefore, if we denote by z_k a function of the space $H^2(\omega)$, satisfying the boundary conditions for $k = 1, K - 1$:

(II.78)
$$\begin{cases}\dfrac{\partial\, z_k}{\partial\, b} = 0\text{ on } \partial\,\omega\,,\\[2ex] z_k = 0\text{ on }\partial\,\omega_j\ \forall\, j = 1\,,\, K\,,\, j \neq k\\[2ex] z_k = 1\text{ on }\partial\,\omega_k\end{cases}$$

Then we define the quantity $\left(\partial_\alpha\, z_k \in H^1_0(\omega)\right)$:

$$\langle q_\alpha\, b_\alpha\,,\, 1\rangle_{\partial\,\omega_k}\ \overset{\text{definition}}{\equiv}\ \int_\omega \operatorname{div} q\, z_k + \left\langle q_\alpha\,,\, \partial_\alpha\, z_k\right\rangle$$

Abusively, this quantity can also be written:

$$\langle q_\alpha\, b_\alpha\,,\, 1\rangle_{\partial\,\omega_k} = \int_{\partial\,\omega_k} q_\alpha\, b_\alpha\ .$$

Let us now introduce a function φ such that:

$$(\text{II.79}) \quad \begin{cases} - \Delta \varphi = - \operatorname{div} q = f_3 \text{ on } \omega , \\[1mm] \varphi = 0 \text{ on } \gamma_{0K} \cup \gamma_{1K}, \quad \varphi = D_k \text{ on } \gamma_{0k} \cup \gamma_{1k} \; \forall \; k = 1 , K - 1, \\[1mm] \dfrac{\partial \varphi}{\partial b} = 0 \text{ on } \gamma_2 , \quad \displaystyle\int_{\gamma_{0k} \cup \gamma_{1k}} \dfrac{\partial \varphi}{\partial b} = \langle q_\alpha b_\alpha , 1 \rangle_{\partial \omega_k} \; \forall \; k = 1 , K - 1 \end{cases}$$

D_k being unknown constants. As a matter of fact, II.79 defines φ as a unique element in the space $H^1 (\omega)$ as soon as f_3 is in $L^2 (\omega)$. This classical exercise is left to the reader. Then we introduce the vector field $q = (q_\alpha)$ by:

$$q = p - \operatorname{grad} \varphi$$

φ being the function defined in II.79. Thus one has for p the following properties:

- $\operatorname{div} p = 0 \text{ on } \omega$, • $p = (p_\alpha) , p_\alpha \in H^{-1} (\omega)$, • $\displaystyle\int_{\partial \omega_k} p_\alpha b_\alpha = 0$

the definition of the last term could be justified as we did for q (see above). Therefore, applying L. Tartar lemma II.3, we conclude that there exists a function ψ unique in the space $L_0^2 (\omega)$ and such that:

$$p = \operatorname{rot} \psi = \left(- \partial_2 \psi , \partial_1 \psi \right)$$

Following the same theoretical approach as we did in section II.4.1, we construct a variational equation, the couple (θ , ψ) is solution to:

$$\text{II.80} \quad \begin{cases} \forall \mu \in W_t , k (\theta , \mu) - \displaystyle\int_\omega \psi \operatorname{curl} \mu - \sum_{i = 1, N} C_i \int_{\gamma_2^i} \mu \, s = - \int_\omega \partial_\alpha \varphi \, \mu_\alpha \\[3mm] \forall \; q \in L_0^2 (\omega), \displaystyle\int_\omega q \operatorname{curl} \theta = 0 , \\[3mm] \forall \; M \in \mathbb{R}^N , M_i \displaystyle\int_{\gamma_2^i} \theta \, s = 0. \end{cases}$$

Apparently this mixed formulation is not well posed, because φ depends on q and therefore on ψ. Hence it cannot be computed previously and then introduced into II.80. Let us show how the g_k functions (see II.77) can be used for decoupling. Let us set:

$$\varphi = \sum_{k=1,K-1} D_k \, g_k + \varphi_0$$

where φ_0 is solution to:

(II.81)
$$\begin{cases} -\,\Delta\,\varphi_0 \,=\, f_3 \text{ on } \omega\,, \\[2mm] \varphi_0 \,=\, 0 \text{ on } \gamma_0 \cup \gamma_1\,, \\[2mm] \dfrac{\partial\,\varphi_0}{\partial\,b} \,=\, 0 \text{ on } \gamma_2 \quad. \end{cases}$$

Then D_k should be such that:

$$D_k \int_{\gamma_{0k} \cup \gamma_{1k}} \frac{\partial\,g_k}{\partial\,b} \,=\, \langle\, q_\alpha\, b_\alpha\,,\,1\rangle_{\partial\,\omega_k} \,-\, \int_{\gamma_{0k} \cup \gamma_{1k}} \partial_\alpha\,\varphi_0\,b_\alpha\,.$$

This condition can be fulfilled as soon as the quantity:

$$\int_{\gamma_{0k} \cup \gamma_{1k}} \frac{\partial\,g_k}{\partial\,b} \quad,$$

is different from zero. But returning to the definition II.77 of g_k, one has:

$$|g_k|^2_{1,\,\omega} \,=\, \int_\omega \partial_\alpha\,g_k\,\partial_\alpha\,g_k \,=\, \int_{\gamma_{0k} \cup \gamma_{1k}} \frac{\partial\,g_k}{\partial\,b}g_k \,=\, \int_{\gamma_{0k} \cup \gamma_{1k}} \frac{\partial\,g_k}{\partial\,b}$$

and therefore :

$$\int_{\gamma_{0k} \cup \gamma_{1k}} \frac{\partial\,g_k}{\partial\,b} \,\neq\, 0 \quad, \text{ otherwise } g_k = 0 \quad,$$

which is impossible because $g_k = 1$ on the same part of the boundary. Furthermore, if θ is the solution to the Kirchhoff-Love model with :

$$\theta_\alpha = - \partial_\alpha u_3, \ u_3 \in V_3^0 \ .$$

As $- \Delta g_k = 0$ on ω , one has :

$$\int_\omega \partial_\alpha g_k \, \theta_\alpha = - \int_\omega \partial_\alpha g_k \, \partial_\alpha u_3 \ + \ \int_{\partial \omega} \frac{\partial g_k}{\partial b} \, u_3 = 0$$

Finally, $(\theta , \psi , C , D) \in W_t \times L_0^2(\omega) \times \mathbb{R}^N \times \mathbb{R}^{K-1}$, is solution to the mixed formulation explicited in II.82.

We set in the following: $M = L_0^2(\omega) \times \mathbb{R}^N \times \mathbb{R}^{K-1}$. This is a more general formulation than the one we obtain at section II.4.1. As a matter of fact, there are $K - 1$ additional unknowns (i.e. the constants D_k for k=1,K-1) and as many equations.

We are going to prove that II.82 admits a unique solution. Basically, we extend the proof given in Theorem II.5 for the simply connected boundary case. As a matter of fact, the proof is very similar, excepted for the determination of D_k .

$$(II.82) \quad \begin{cases} \forall\, \mu \in W_t, \; k(\theta, \mu) - \int_\omega \psi \, \text{curl}\, \mu - \sum_{i=1,N} C_i \int_{\gamma_1^i} \mu_s \\ \qquad + \sum_{k=1,K-1} D_k \int_\omega \partial_\alpha g_k \, \mu_\alpha = - \int_\omega \partial_\alpha \varphi_0 \, \mu \\[2ex] \forall\, p \in L_0^2(\omega), \; \int_\omega p \, \text{curl}\, \theta = 0, \\[2ex] \forall\, M = (M_i) \in \mathbb{R}^N, \; M_i \int_{\gamma_2^i} \theta_s = 0, \\[2ex] \forall\, L = (L_i) \in \mathbb{R}^{K-1}, \; L_i \int_\omega \partial_\alpha g_k \, \theta_\alpha = 0. \end{cases}$$

Theorem II.7
Let us set the mixed formulation:

$$\begin{cases} \text{find } (\theta, \Lambda) \in W_t \times M \quad \text{such that :} \\ \forall\, \mu \in W_t, \; k(\theta, \mu) + b(\mu, \Lambda) = g(\mu) \\ \forall\, \Xi \in M, \; b(\Xi, \theta) = 0 \end{cases}$$

where W_t is defined in II.39, and the bilinear form $k(.,.)$ is given in II.37. The bilinear form is defined in the space $W_t \times M$ by:

$$\begin{cases} (\mu, \Lambda) \in W_t \times M, \; \Lambda = (p, C, D) \\ b(\theta, \Lambda) = - \int_\omega p \, \text{curl}\, \mu - \sum_{i=1,N} C_i \int_{\gamma_2^i} \mu_s + \sum_{k=1,K-1} D_k \int_\omega \partial_\alpha g_k \, \mu_\alpha \end{cases}$$

Finally the linear form $g(.)$ is:

$$\forall \; \mu \in W_t, \; g\left(\mu\right) = - \int_\omega \partial_\alpha \varphi_0 \, \mu_\alpha$$

Then the previous mixed formulation has a unique solution.   ∎

Sketch of the proof

Because of its similarity with the proof of theorem II.5, we only focus on the new points. This is the so-called inf-sup condition, and particularly concerning the constant $D \in \mathbb{R}^{K-1}$. We already know that for any $\Lambda = (\psi, C, D)$ in M :

$$(\text{II.83}) \quad \sup_{\mu \in W_t} \frac{b\left(\mu, \Lambda\right)}{\|\mu\|_{W_t}} \geq c_0 \left[\|\psi\|_{L_0^2(\omega)} + \sum_i |C_i|\right] - c_1 \left[\sum_{i=1, K-1} |D_i|\right]$$

where c_0 and c_1 are two constants and $\Lambda = (\psi, C, D)$; $C = (C_i), D = (D_i)$. Let us then consider an element – say z_k – of the space $H^2(\omega)$ as it has been defined in II.78. We set:

$$\mu_\alpha = - \sum_{k=1, K-1} D_k \, \partial_\alpha z_k$$

and, because of the definition of z_k , one has :

$$b\left(\mu, \Lambda\right) = \sum_{\substack{k=1, K-1 \\ j=1, K-1}} D_k D_j \int_\omega \partial_\alpha g_k \, \partial_\alpha z_j$$

or else (because $- \Delta g_k = 0$):

$$b\left(\mu, \Lambda\right) = \sum_{\substack{k=1, K-1 \\ j=1, K-1}} D_k D_j \int_{\partial\omega} \frac{\partial g_k}{\partial b} z_j$$

But we know that z_j is not zero only on the component $\partial \omega_j$ of $\partial \omega$ and that $\dfrac{\partial g_k}{\partial b}$ is zero on the free edge of $\partial \omega_j$. Thus we obtain:

$$b\left(\mu, \Lambda\right) = \sum_{\substack{k=1, K-1 \\ j=1, K-1}} D_k D_j \int_{\gamma_{0j} \cup \gamma_{1j}} \frac{\partial g_k}{\partial b}$$

From the definition of g_k and g_j (see II.77), it clearly appears that:

$$b(\mu, \Lambda) = \sum_{\substack{k=1, K-1 \\ j=1, K-1}} D_k D_j \int_\omega \partial_\alpha g_k \partial_\alpha g_i$$

and finally, with the same choice for μ as before :

$$b(\mu, \Lambda) \geq c_0 \sum_{k=1, K-1} |D_k|^2$$

where c_0 is the smallest eigenvalue of the positive definite matrix $\int_\omega \partial_\alpha g_k \partial_\alpha g_i$

Noticing that :

$$||\mu||_{W_t} \leq \sum_{k=1}^{K-1} |D_k| \, ||z_k||_{2,\omega} \leq c_1 \left[\sum_{k=1}^{K-1} |D_k|^2 \right]^{1/2}$$

where c_1 is a constant which is independent of D_k, we conclude to the estimate:

$$(II.84) \quad \sup_{\mu \in W_t} \frac{b(\mu, \Lambda)}{||\mu||_{W_t}} \geq \frac{c_0}{c_1} \left[\sum_{k=1, K-1} |D_k|^2 \right]^{1/2} \quad .$$

Coupling II.84 and II.83 enables one to derive the so-called inf-sup condition:

$$\sup_{\mu \in W_t} \frac{b(\mu, \Lambda)}{||\mu||_{W_t}} \geq C \, ||\Lambda||_M$$

where $\Lambda = (\psi, C, D) \in M$, and $M = L_0^2(\omega) \times \mathbb{R}^N \times \mathbb{R}$. The following of the proof of Theorem II.7 is identical to the one given for Theorem II.5. ■

II.4.8.2 *Characterization of* u_3

Once φ, ψ and θ have been computed, u_3 can be obtained by solving the equation:

$$(II.85) \quad \begin{cases} - \Delta u_3 = \operatorname{div} \theta = \partial_\alpha \theta_\alpha \text{ on } \omega, \\[2mm] u_3 = 0 \text{ on } \gamma_0 \cup \gamma_1, \\[2mm] \dfrac{\partial u_3}{\partial b} = - \theta_\alpha b_\alpha \text{ on } \gamma_2. \end{cases}$$

which is exactly the same system as the one we derived in section II.4.3, formula II.69). Obviously, the existence and uniqueness of a solution to II.85 is classical. It is worth noticing that the connected components of the boundary $\partial \omega$ of ω do not play any role in the characterization of u_3, except in the effective computation of θ which is considered as a right-hand side in II.85.

II.4.8.3 Interpretation of the solved model

Even if the derivation of the mixed formulation II.81 is rigorous, we assumed – as usual – a regularity of the unknown functions, in order to be authorized to use the Stokes (as a matter of fact Green version) formula. It is therefore necessary to interpret the solution to the mixed formulation II.81 – II.85.

First of all, equations II.82 lead to:

$$- \partial_\beta m_{\alpha \beta} + (\operatorname{rot} \psi)_\alpha + \partial_\alpha \varphi = 0 \text{ on } \omega,$$

where :

$$m_{\alpha \beta} = \frac{2 E \varepsilon^3}{3 (1 - \nu^2)} \left\{ (1 - \nu) \gamma_{\alpha \beta}(\theta) + \nu \gamma_{\mu \mu}(\theta) \delta_{\alpha \beta} \right\},$$

$$\varphi = \varphi_0 + \sum_{k = 1, K - 1} D_k g_k$$

and φ_0 is the unique solution to the equation II.81. Furthermore, the boundary conditions satisfied by the variational solution are:

$$\begin{cases} \theta_\alpha = 0 \text{ on } \gamma_0, \\ \theta_s = \theta_\alpha a_\alpha = 0 \text{ on } \gamma_1, \end{cases}$$

$$\begin{cases} m_{\alpha\beta}\, b_\alpha\, b_\beta = 0 \ \text{on} \ \gamma_1 \cup \gamma_2, \\ m_{\alpha\beta}\, a_\alpha\, b_\beta + \psi - C_i = 0 \ \text{on} \ \gamma_2^i \,, \ i = 1\,, N \end{cases}$$

The last of these equations leads to:

$$\partial_s \left(m_{\alpha\beta}\, a_\alpha\, b_\beta \right) + \partial_s \psi = 0 \ \text{on} \ \gamma_2^i \,, \ i = 1\,, N$$

and then from:

$$- \partial_\beta\, m_{\alpha\beta}\, b_\beta + (\text{rot}\ \psi)_\alpha\, b_\alpha + \frac{\partial\varphi}{\partial b} = 0 \ \text{on} \ \gamma_2$$

and noticing that $\dfrac{\partial\varphi}{\partial b} = 0$ on γ_2, we deduce that:

$$(\text{rot}\ \psi)_\alpha\, b_\alpha = - \partial_2\, \psi\, b_1 + \partial_1\, \psi\, b_2 = \partial_1\, \psi\, a_1 + \partial_2\, \psi\, a_2$$
$$= \partial_s\, \psi = \partial_\beta\, m_{\alpha\beta}\, b_\alpha$$

Finally:

$$\partial_s \left(m_{\alpha\beta}\, a_\alpha\, b_\beta \right) + \partial_\beta\, m_{\alpha\beta}\, b_\alpha = 0 \ \text{on} \ \gamma_2 \ .$$

The last main step is to prove that $\theta_\alpha = - \partial_\alpha\, u_3$. In order to establish this relation, let us recall that from the mixed system II.82, one has:

(II.86)
$$\begin{cases} \text{curl}\ \theta = 0 \ \text{on} \ \omega \\[2mm] \displaystyle\int_{\gamma_2^i} \theta_s = 0 \ \ i = 1\,, N \\[2mm] \displaystyle\int_\omega \partial_\alpha\, g_k\, \theta_\alpha = 0 \ \ k = 1\,, K-1 \ . \end{cases}$$

But from the boundary conditions concerning θ (space W_t), one has:

$$\int_{\partial \omega_k} \theta_s = 0 \quad \forall \ k = 1, K \ .$$

Thus from Lemma II.2, one can ensure that there exists a function, say z, such that:

$$\begin{cases} z = - \text{grad } z \ , \ z \in H^2(\omega) \, , \dfrac{\partial z}{\partial b} = 0 \text{ on } \gamma_0 \\[3mm] z = 0 \text{ on } \gamma_{0K} \cup \gamma_{1K} , \ z = C_k \text{ on } \gamma_{0k} \cup \gamma_{1k} , \ \forall \ k = 1, K-1 \ . \end{cases}$$

But from the last relation II.86, one has for $k = 1, K - 1$:

$$0 = \int_{\omega} \partial_\alpha g_k \, \partial_\alpha z = \int_{\partial \omega} \frac{\partial g_k}{\partial b} z = \sum_{j=1, K-1} C \int_{\gamma_{0j} \cup \gamma_{1j}} \frac{\partial g_k}{\partial b} g_j$$

$$= \sum_{j=1, K-1} C_j \int_{\omega} \partial_\alpha g_k \, \partial_\alpha g_j \ .$$

This implies (see the proof of Theorem II.7) that:

$$C_j = 0 \quad \text{for } j = 1, K-1$$

Thus:

$$\begin{cases} \theta = - \text{grad } z \\ z \in V_3^0 \end{cases}$$

It is then easy to check first that z is solution to the Kirchhoff-Love model and then (this is really the last step), that z is also the unique solution to II.85. Let us emphasize the consequence of this last remark. Usually the solution of a Laplace equation lies in the space $H^1(\omega)$. Only if a regularity assumption is satisfied, then z is in $H^2(\omega)$. As a matter of fact, we know from the previous result that z is in $H^2(\omega)$. Hence the suitable regularity assumptions are necessarily satisfied by the functions θ_α solution of the mixed formulation II.82. Finally, we conclude that we really solved the right problem.

II.5   A comparison between the mixed method and the one of section II.2.4

Let us consider a simple case where the boundary of the plate is simply fixed in the sense that;

$$\gamma = \gamma_1.$$

Furthermore we suppose that the boundary γ is piecewise linear (case 1 in section II.2.4).

Then the bending plate model is the following one:

find u_3 such that:

$$\frac{2\,E\,\varepsilon^3}{3\,(1 - v^2)}\,\Delta^2 u_3 = f \quad \text{on } \omega$$
$$u_3 = 0, \quad \Delta u_3 = 0 \quad \text{on } \gamma$$

We set :

$$- \frac{2\,E\,\varepsilon^3}{3\,(1 - v^2)}\,\Delta u_3 = \varphi \quad \text{on } \omega \text{ and } u_3 = 0 \quad \text{on } \gamma$$

then φ is solution to the following problem:

$$- \Delta\varphi = f \quad \text{on } \omega, \quad \varphi = 0 \quad \text{on } \gamma$$

which is precisely the same function as the one used in the mixed formulation in the expression of the transverse shear stress. But if we remember that precisely the resultant transverse shear stress is defined by:

$$q_\alpha = \partial_\beta\, m_{\alpha\beta} = - \frac{2\,E\,\varepsilon^3}{3\,(1 - v^2)}\,\partial_\alpha\,\Delta\,u_3 = \partial_\alpha\,\varphi ,$$

and therefore one has in this case $\psi = 0$. But obviously one has to be careful that this is a very particular case. As a matter of fact the function ψ is very much connected to the boundary conditions satisfied by the plate.

REFERENCES

[1] BREZIS H., [1983], Analyse fonctionnelle - Théorie et applications - Masson, Paris.

[2] GRISVARD P., [1975], Behavior of the solutions of an elliptic boundary value problem in a polygonal or polyhedral domain; Numerical solution of partial differential equations III, (SYNSPADE), p. 207-274, Academic Press New York

[3] WILLIAMS M. L., [1961], The bending stress distribution at the base of a stationary crack, Journal of Applied Mechanics; March 1961.

[4] GLOWINSKI R., [1973], Approximations externes, par éléments finis de Lagrange d'ordre un et deux, du problème de Dirichlet pour l'opérateur biharmonique - Méthodes itératives de résolution des problèmes approchés, in Topics in Numerical Analysis (J.J.H. Miller, editor), p. 123-171, Academic Press, London.

[5] DESTUYNDER Ph., NEVERS Th., [1988], Une modification du modèle de Mindlin pour les plaques minces en flexion présentant un bord libre, Math. Modelling Numer. Anal. 22, p. 217-242.

[6] REISSNER E., [1945], The effect of transverse shear deformations on the bending of elastic plates, J. Appl. Mech. 12, A69-A77.

[7] MINDLIN R., [1951], Influence of rotatory inertia and shear on flexural motions of isotropic elastic plates, J. Appl. Mech. 18, p. 31-38.

[8] NAGHDI P. M., [1972], Handbuch der Physik, Band V/a/2, p. 425-640, , Springer Verlag, Berlin.

[9] LADEVEZE P., [1988], Les modèles classiques et leurs extensions pour le calcul des plaques - Calcul des structures et intelligence artificielle n° 2, Pluralis, Paris.

[10] DESTUYNDER Ph., [1980], Une théorie asymptotique des plaques minces en élasticité linéaire. Doctoral dissertation Univ. P.M. Curie, Paris 6.

[11] MATHUNA D. O., [1989], Mechanics, Boundary Layers and function spaces - Birkhauser, Boston.

[12] DUVAUT G., LIONS J. L., [1972], Les inéquations en mécanique et en physique, Dunod, Paris.

[13] TEMAM R., [1979], Navier Stokes equations - North Holland Studies in Mathematics and its applications n° 2, Amsterdam.

[14] GIRAULT V., RAVIART P. A., [1986], Finite element methods for Navier-Stokes equations, S.C.M. n° 5, Springer Verlag - Berlin.

[15] BABUSKA I., [1971], Error bounds for finite element methods, Numer. Math., 16, p. 323-333.

[16] LIONS J. L., [1973], Perturbations singulières dans les problèmes aux limites et en contrôle optimal - Lectures Notes in Mathematics, Vol. 323, Springer Verlag, Berlin.

[17] DESTUYNDER Ph., NEVERS Th., [1988], A new finite element scheme for bending plates, Comp. Meth. Appl. Mechs Eng, 68, p. 127-139.

[18] DESTUYNDER Ph., NEVERS Th., [1990], Some numerical aspects of mixed finite elements for bending plates, Comp. Meth. Appl. Mechs. Eng. 78, p. 73-87.

[19] BERCOVIER M., [1978], Perturbation of mixed variational problems - Application to mixed finite element methods, RAIRO, Anal. Numer. 12, p. 211-236.

[20] BREZZI F., [1974], On the existence, uniqueness and approximation of saddle point problems arising from Lagrangian Multipliers, RAIRO, R2, p. 129-151.

[21] BREZZI F., FORTIN M., [1986], Numerical Approximation of Mindlin-Reissner plates, Math. Comp. 47 (175) p. 151-158.

Chapter 3

FINITE ELEMENT APPROXIMATIONS FOR SEVERAL PLATE MODELS

III.0 A summary of the chapter

After a brief reminder of finite elements methods we discuss the possible approximations of the penalty or modified penalty model which has been studied in Chapter II. The connections with the famous QUAD 4 element (and its numerous variants) are examined. This permits one to give a partial justification of this well-known element and an extension to general cases including triangles. Then the natural duality technique is used to construct a new kind of structural finite elements which are analyzed from the error point of view.

III.1 Basic results in finite element approximation

III.1.1   Several useful definitions

Let us start with basic definitions that we use continuously in this chapter. They are quite classical. The first presentation of these basic aspects is due to G. Strang and G. Fix [1]. Then substantial improvements were given by P.G. Ciarlet [2], and P.A. Raviart-J.M. Thomas [3]. Hence nothing is new in this section. But a brief recall may be convenient for the reader. Proofs and details are omitted, for the sake of brevity. First of all, let us give the definition of a mesh.

Definition III.1
A two-dimensional mesh T^h *for an open set* ω, *is a collection of elementary triangles or quadrilaterals, say K, such that:*

$\forall\ K_1,\ K_2 \in T^h$ *one and only one of the four following properties is true:*

i) $K_1 = K_2$
ii) $K_1 \cap K_2$ *is a common side*
iii) $K_1 \cap K_2$ *is a common vertex*
iv) $K_1 \cap K_2$ *is empty.*

In addition, a mesh T^h *of* ω *is such that:*

$$\cup K \; = \; \bar{\omega} (closure \; of \; \omega)$$
$$K \in T^h$$

This implies (in our context) that $\bar{\omega}$ has a polygonal boundary. In practical application the boundary of the mesh is only an aproximation of the one of ω.   ■

Definition III.2

A Lagrange finite element is a triplet $\left(K, \; \Sigma_K, \; P \right)$ where K is a triangle or a quadrilateral, Σ_K is a set of nodes on K (for instance the three summits), and P is a space of polynomials defined on K (the polynomials of degree 1 or 2 will be used in the following).   ■

A basic property of a Lagrange finite element is unisolvency.

Definition III.3.

"Unisolvent element". A Lagrange finite element is unisolvent if for any set of n scalar values (n being the number of nodes), there exists a unique element, in the space P, which takes these values at the nodes of K.   ■

The finite elements used later on are all unisolvent. Let us mention them hereafter.

1. The P_1 – Lagrange

Figure III.1

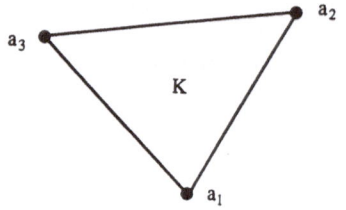

K is a triangle

$\Sigma \; = \; \{a_1 \, , \, a_2 \, , \, a_3\}$ (three vertices)

$P = P_1$ (first degree polynomials)

2. The Q_1 – Lagrange

Figure III.2

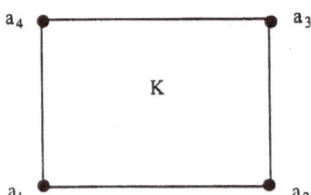

K is a square or a rectangle

$\Sigma \; = \; \{a_1 \, , \, a_2 \, , \, a_3 \, , \, a_4\}$ (four vertices)

$P = Q_1$ (first degree polynomials with respect to each variable)

3. The deformed Q_1 – Lagrange

Figure III.3

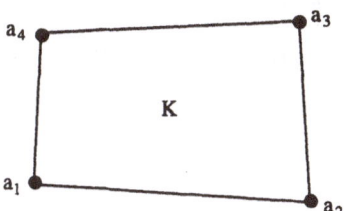

K is a four-side plane figure

Σ_K is the set of the four vertices

This element is much more complicated than most finite element books suggest. A special interest for this strange element in plate modelling, is certainly due to the success of the famous QUAD4 element which was introduced by MacNeal [4] and T. Hughes [5]. The element K is a convex quadrilateral which is the image of a square (see Figure III.3) through a bilinear component mapping F_K (defined by the coordinates of the four summits a_1, a_2, a_3 and a_4 because the Q_1 – Lagrange is unisolvent). The mapping F_K is not linear except along straight lines parallel to the side of the reference square. Hence the functions of the space P which are defined by the image of polynomials of Q_1 by the mapping F_K, are more complicated. In other words:

$$P = \left\{ p \,|\, p = q \circ F_K^{-1}, \; q \in Q_1 \text{ therefore q is defined on } \widehat{K} \right\} .$$

Hence P contains rational functions and therefore can be singular at the roots of the denominator. As a matter of fact they correspond to the intersections of the opposite sides of the element K. Another important point is that the origin of the reference square is mapped onto the intersection point of the median lines (see Figure III.3). This is different from the intersection of the diagonal lines except in very particular cases (rectangles or diamond-shaped elements).

4. The P_2 – Lagrange

Figure III.4

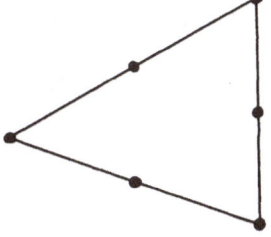

K is a triangle

Σ_K is the set of the three summits and the three mid-points on each side of the triangle

$P = P_2$ (second-degree polynomials)

5. The Q_2 – Lagrange

Figure III.5

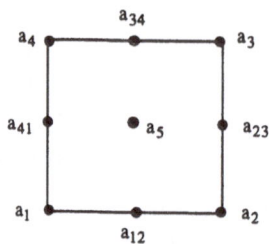

K is a rectangle

Σ_K is the set of the four summits, the four mid-points on the sides of K and the center of gravity

$P = Q_2$ (space of second-degree polynomials with respect to each variable).

6. The incomplete Q_2 – Lagrange

Figure III.6

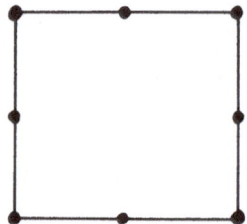

This element is analogous to the former one except that node a_5 is cancelled (see Figure III.5) and the bubble function is also withdrawn from the space Q_2. This function is equal to the product of the four side equations of the rectangle K. It is denoted by B_K (the bubble function). Hence we set:

$$Q_{2I} = Q_2 - \{B\}$$

7. The deformed incomplete or complete Q_2 – Lagrange

Figure III.7

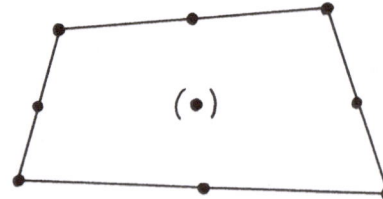

In this case the element K is still an arbitrary quadrilateral. The nodes are the same as the ones introduced for the Q_2 or incomplete Q_2 – Lagrange. We refer the reader to the comments on the Q_1 – Lagrange.

8. The mini-element

Several finite element schemes – that we use later on – are low order but require an internal degree of freedom. There are two possibilities for adding such "a facility". The first one is the bubble method. The second one is the Ramses technique.

P_1 + Bubble element. This is a triangular element (see Figure III.1), where we add to Σ_K the center of gravity and to P, the bubble function which is defined as the product of the three side equations (normalized by adjusting the value to one at the center of gravity). Obviously the bubble function is a third order polynomial.

P_1 + Ramses element. This is the same element as above except that the bubble function is replaced by the Ramses function. It is defined as follows: the triangle K is divided into three subtriangles based on the center of gravity (see Figure III.8), and the function is:

$$\text{Ramses} = \{\text{Ramses}|_{K_i} \in P_1 \quad \forall \; i = 1, 2, 3,$$
$$\text{Ramses} \in C^0(K), \; \text{Ramses}(a_i) = 0, \; i = 1, 2, 3$$
$$\text{Ramses}(a_4) = 1\}$$

Figure III.8

$$K = K_1 \cup K_2 \cup K_3$$
$$\Sigma_K = \{a_1, a_2, a_3, a_4\}$$
$$P = P_1 \oplus \{\text{Ramses}\}$$

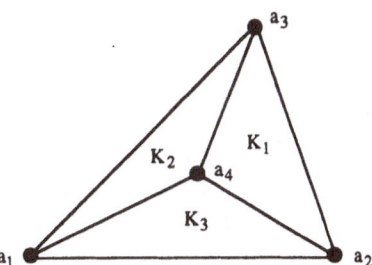

Q_1 + Bubble or Ramses element. This element is based on a rectangle. We add the center of gravity to the set Σ_K on Figure III.2 and the bubble function (defined as the product of the four side equations normalized by the value one at the center of gravity) or the Ramses function to the space P. The Ramses function for rectangles is defined as follows (see Figure III.9) :

$$\text{Ramses} = \{\text{Ramses}|_{K_i} \in Q_1 \text{ for } i \in \{1, 2, 3, 4\}$$
$$\text{Ramses}(a_i) = 0 \text{ for } i \in \{1, 2, 3, 4\} \; ; \; \text{Ramses}(a_5) = 1\}$$

Figure III.9

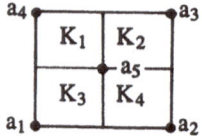

III.1.2   A brief recall concerning error estimates

Let us begin by defining the finite element spaces which can be generated with the elements introduced in the previous section. For sake of clarity let us emphasize the case of the P_1, P_2, Q_1 or Q_2 elements. The reader will generalize easily with other elements. Anyway this will be discussed ahead.

III.1.2.1   Spaces $P_1 - Q_1$ or $P_2 - Q_2$.

Let T^h be a mesh of an open set ω (see Definition III.1). Then we set for $k = 1$ or 2:

$$(III.1) \quad V_k^h = \left\{ v \mid \in H^1(\omega) \cap C^0(\bar{\omega}); \ \forall \ K \in F^h, \ v_{|K} \in P_k; \ \forall \ K \in Q^h, \ v_{|K} \in Q_k \right\}$$

where F^h (respectively Q^h) is the collection of triangles (respectively rectangles) of T^h.

Then we assume that we have a family of meshes with respect to the index h, which also denotes the generalized mesh size. This family is said to be regular, following the terminology of Ciarlet [2] or Raviart-Thomas [3], if there exists a constant – say c – such that for any element of T^h and for any T^h one has:

$$0 < c \leq \frac{\rho}{h}$$

where h is the largest side of a triangle or a quadrilateral and ρ the diameter of the largest circle contained in a triangle or in a rectangle. This basic property has to be modified for deformed Q_1 elements as follows. Let K_1 and K_2 be two subtriangles constructed from an arbitrary *convex* quadrilateral using the diagonal (there are two possibilities). The above condition, concerning the quadrilaterals of T^h, becomes:

$$\forall \ K \in Q^h, \ \forall \ K_i \text{ subtriangle of } K: \ 0 < c \leq \frac{\rho_i}{h}$$

where ρ_i is the diameter of the largest circle contained in K_i (there are four possible subtriangles for any quadrilateral, depending on the choice of the diagonal).

Assuming this regularity assumption, one has the following error estimates due to G. Strang and G. Fix [1].

Theorem III.1.

Let u *be a function of the space* $H^{k+1}(\omega)$ *for k=1 or 2, and let* πu *be the linear interpolate of* u *in the space* V_k^h *(defined at III.1), such that:* $\pi u(a_i) = u(a_i) \; \forall \; a_i$, *node of the mesh* T^h *(the nodes are the union of the points of* Σ_K *for* $K \in T^h$, *or more precisely the summits of* K *for linear elements and additionally the mid-points of each side of the elements* $K \in T^h$, *for k=2). Then* πu *is uniquely defined,* π *is linear and continuous from* $H^1(\omega) \cap C^0(\omega)$ *into the space* $H^m(\omega)$ *with* m = 0,1, *and there exists a constant C, independent of the mesh size as soon as the mesh family* T^h *is regular, such that:*

$$\|u - \pi u\|_{m,\omega} \le C h^{k+1-m} |u|_{k+1,\omega} \; .$$

In particular for k = 1 *(space* $P_1 - Q_1$ *):*

(III.2)
$$\begin{cases} \|u - \pi u\|_{0,\omega} \le c h^2 |u|_{2,\omega} \\ \|u - \pi u\|_{1,\omega} \le c h |u|_{2,\omega} \end{cases},$$

and for k = 2 *(space* $P_2 - Q_2$*):*

(III.3)
$$\begin{cases} \|u - \pi u\|_{0,\omega} \le C h^3 |u|_{3,\omega} \\ \|u - \pi u\|_{1,\omega} \le C h^2 |u|_{3,\omega} \end{cases}.$$

■

This result is very often used in the following. The proof is not necessary for the understanding of the following. But it can be interesting for the reader to catch its technical details. The main references are G. Strang - G. Fix [1], P.A. Raviart - J.M. Thomas [3], and P.G. Ciarlet [2].

Two other error estimate results will also be used. They are not so well known in reference books. But most numerical analysts working in finite elements know them.

Theorem III.2

Let us consider the same assumption as in Theorem III.1 *concerning the meshes* T^h. *We define the projection of an element* u *lying in the space* $H^1(\omega)$ *by:* P u $\in V_k^h$ *such that:*

$$\|u - Pu\|_{0,\omega} = \inf_{v \in V_k^h} \|u - v\|_{0,\omega} \; .$$

Then there exists a constant c independent of the mesh size and such that :

$$\|u - Pu\|_{0,\omega} \le c\,h\|u\|_{1,\omega} \quad \text{for} \quad k = 1 \text{ or } 2 \; . \qquad \blacksquare$$

Remark III.1

This result seems to be very simple, but it is not because only the norm $H^1(\omega)$ appears at the right-hand side of the previous inequality instead of $H^2(\omega)$ or $H^3(\omega)$ as it was the case in Theorem III.1. As a matter of fact the interpolate π (P_1 - Q_1 Lagrange) is not defined from $H^1(\omega)$ into $L^2(\omega)$ because in two (or more) dimensions: $H^1(\omega) \not\subset C^0(\overline{\omega})$. Hence it is necessary to use the projection P. The proof of Theorem III.2 can be obtained from Theorem III.1 by applying an interpolation between the operator I - P considered as an element of the space $L(L^2(\omega), L^2(\omega))$, and the operator I-P, but considered as an element of the space $L(L^2(\omega), H^2(\omega))$. The details of the proof can be found in the book of D. Huet [6] or the one of J.L. Lions - E. Magenes [7]. $\qquad \blacksquare$

us also give a simple extension of Theorem III.2. Let us introduce the finite dimensional space V_k^{hd} such that:

$$V_k^{hd} = \left\{ v \in L^2(\omega) \,/\, \forall\, K \in T^h, \; v_{|K} \in P_k \text{ if K is a triangle} \,;\, v_{|K} \in Q_k \text{ if K is a quadrilateral} \right\}$$

The functions of V_k^{hd} are discontinuous. Then we define the projection of an element $u \in H^1(\omega)$ onto V_k^{hd} by:

$$P^d u \in V_k^{hd} \; ,$$

$$\|u - P^d u\|_{0,\omega} = \inf_{v \in V_k^{hd}} \|u - v\|_{0,\omega} \; .$$

Thus, noticing that $P\,u$, which lies in the space V_k^h, is also in V_k^{hd}, we deduce that:

$$\|u - P^d u\|_{0,\omega} \le c\,h\|u\|_{1,\omega} \quad \text{for} \quad k = 1 \text{ or } 2 \; .$$

The last result that we use concerns error estimates in "negative" Sobolev spaces.

Theorem III.3

Let u *be an element of the space* $H^1(\omega)$ *and we assume that the assumptions of Theorem III.1 are still satisfied (concerning the meshes* T^h *). Then there exists a constant c such that:*

$$\|u - Pu\|_{-1,\omega} \le c\,h^2\|u\|_{1,\omega} \quad \text{for} \quad k = 1 \text{ or } 2$$

where P u *is the projection of* u *onto* V_k^h *defined at* Theorem III.2.   ∎

The proof is very classical and could be omitted. But, because of its simplicity and its importance in the rest of the chapter, we sketch it hereafter.

Proof of Theorem III.3
First of all, by definition of the norm in the space $H^{-1}(\omega)$, dual of $H_0^1(\omega)$, one has :

$$
\| u - Pu \|_{-1,\omega} = \sup_{\varphi \in H_0^1(\omega)} \frac{\int_\omega \varphi(u - Pu)}{\| \varphi \|_{1,\omega}} = \sup_{\varphi \in H_0^1(\omega)} \frac{\int_\omega (\varphi - P\varphi)(u - Pu)}{\| \varphi \|_{1,\omega}} ,
$$

because the definition of the projection of u onto V_k^h is such that :

$$
\forall\ v \in V_k^h \quad \int_\omega v(u - Pu) = 0 .
$$

Hence from Schwarz inequality and Theorem III.2, we deduce Theorem III.3.   ∎

Remark III.2
When the deformed Q_1 or Q_2 elements are used in the definition of the approximation space, the previous results are still true but the proof needs a control on the deformation of the quadrilateral. One possibility consists in prescribing the condition, mentioned previously :

$$
\forall\ K \in Q^h : 0 < c \le \frac{\rho_i}{h} ,
$$

for all subtriangle K_i of the quadrilateral K defined by the diagonal. But one can prove that this conditions is too restrictive. Nevertheless it is almost necessary in order to avoid ill-conditioning of the system arising from the approximate model.   ∎

Remark III.3
The mini-element is richer than the P_1 or Q_1 element because it contains an internal degree of freedom. But although it is convenient for technical reasons in the following, it does not enable one to improve the accuracy of the error estimate given in Theorem III.1 where the interpolate πu of u is defined without these degrees of freedom. Nevertheless the polynomial invariance is still satisfied and this enables one to apply the classical error estimates of P. G. Ciarlet and P. A. Raviart. Conversely, the incomplete Q_2 element implies the loss of the benefit of second-order polynomials. Hence the error estimate is limited to the one obtained with Q_1 elements.   ∎

III.2   C¹ elements

The functions of the space V_k^h defined in the previous section are continuous. It is easy to check it. But they are not C¹.

Figure III.10   Argyris element

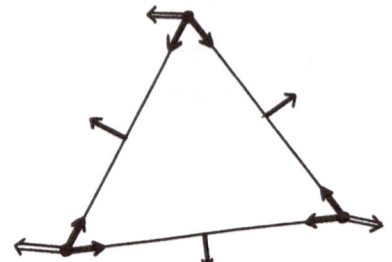

/Degrees of freedom
| Value at a point
| →   Value of a derivate in the direction of the arrow
\ ⇒   Three values of the second-order derivatives

K is a triangle
In order to solve fourth-order problems like the plate bending model, several authors have suggested to use C¹ elements. The most popular of them is the Argyris element (J. Argyris [8]). For defining such an element, it is necessary to use fifth-degree polynomials. Furthermore, the degrees of freedom of the element not only involve the values of a function at several nodes but also the first and second order derivatives. For further details we refer to the book by Ciarlet [2] and a comprehensive analysis of such elements is given in the work published by M. Bernadou [9]. The Argyris element is summarized on Figure III.10. Let us mention (cf. Ciarlet [2]) that the error estimate between a function u lying in the space $H^6(\omega)$ and its interpolation (defined by the values at the degrees of freedom of the element and belonging to the space P_5 (fifth-degree polynomials) on each element), is :

$$\|u - \pi u\|_{m,\omega} \le c\,h^{6-m}|u|_{6,\omega} \text{ for } m = 0,1,2.$$

III.3   Primal finite element methods for bending plates

Let us recall that we define by primal finite element methods, methods in which only the displacement is approximated.

Then, the goal of this short paragraph is just to give an idea of the most classical methods used in engineering applications. The presentation is neither complete nor fully convincing. May be because we are not convinced ourselves ! The first method uses the J. Argyris element [10]. Similar elements are extensively described in the book of Ciarlet [2]. The extension to shells based on the Koiter model has been performed by M. Bernadou in his Doctoral Dissertation [9]. J. Argyris recommended also the TRIM element based on the natural formulation or the QUAD family (QUAD 4 - QUAD 8 - QUAD 12) (see Parisch [11]).

The mesh consisting of triangles, let us set first :

$$V_3^{0h} = \left\{ v \mid v \in V_3^0, \ \forall \ K \in T^h, \ v_{|K} \in P_5 \right\}^1$$

where :

$$V_3^0 = \left\{ v \mid v \in H^2(\omega), \ v = 0 \text{ on } \gamma_0 \cup \gamma_1, \ \frac{\partial v}{\partial b} = 0 \text{ on } \gamma_0 \right\}.$$

$V_3^0 \subset H^2(\omega)$, hence the elements of V_3^{0h} must be $C^1(\overline{\omega})$. Then the approximate model is defined by:

(III.4)
$$\begin{cases} \text{find } u_3^h \in V_3^{0h} \text{ such that:} \\ \forall \ v \ \in \ V_3^{0h}, \ a\left(u_3^h, v\right) = I(v) \end{cases}$$

where:

$$\begin{cases} a(u_3, v) = \dfrac{2 E \epsilon^3}{3(1 - v^2)} \displaystyle\int_\omega (1 - v) \partial_{\alpha\beta} u_3 \, \partial_{\alpha\beta} v + v \Delta u_3 \Delta v, \\[2mm] I(v) = \displaystyle\int_\omega f_3 \, v \end{cases}$$

The existence and uniqueness of a solution to (III.4) are obvious because $V_3^{0h} \subset V_3^0$. Furthermore one has the classical error estimate:

$$\left\| u_3 - u_3^h \right\|_{2,\omega} \le c \, h^4 |u|_{6,\omega}$$

where u_3 is solution to the continuous Kirchhoff-Love model (see chapter II):

$$\begin{cases} u_3 \in V_3^0 \\ \forall \ v \in \ V_3^0, \ a(u_3, v) = I(v) \end{cases}.$$

The second class of primal methods that we mention is certainly due to Irons [12] and is known as the class of non-conforming methods. The methods are presented in the book of O.C. Zienkiewicz [13] and a mathematical analysis has been carried out by P. Lesaint and P. Lascaux [14]. For a long time, this element was the only one available in the famous finite element code, EL. FINIS, developed by Dassault-Breguet company. Let us point out that a new revival of non-conforming methods has recently appeared with the works of D. Arnold and R. Falk, (see also an interesting comparison between the 3 D and Reissner-Mindlin model by D. Arnold and R. Falk [15]). The basic idea is to use an approximation space V_3^{0h} which is not included in V_3^0 (the continuity of the first order derivatives is partially relaxed). But it is necessary to extend the

---

[1] The Argyris element enables one to construct this space!

definition of the bilinear form a (. ,) in order to use it on the space V_3^{0h}. Irons' brilliant idea was to set for any u_3 and v in V_3

$$a^h(u_3, v) = \sum_{K \in T^h} \frac{2 E \varepsilon^3}{3(1 - v^2)} \int_K (1 - v) \partial_{\alpha\beta} u_3 \, \partial_{\alpha\beta} v + v \, \Delta u_3 \, \Delta v.$$

One should notice that:

$$\forall \, u_3, v \in V_3^0, \quad a(u_3, v) = a^h(u_3, v)$$

but this equality is no more true if u_3 and v are not $C^1(\overline{\omega})$. Two possible choices for V_3^{0h} are obtained with only cubic polynomials (instead of fifth degree for C^1 elements). The triangle version has nine (or ten) degrees of freedom and the rectangle version (due to Adini [16]) has 12 degrees of freedom. They are represented on Figure III.12.
The numerical convergence of such non-conforming methods has been first established by Irons using the so-called Patch-test. This convergence criterion states that the normal components of the stresses, which are involved in the plate bending model, are continuous in average value. Then error estimates using the h-dependent norm:

$$u \in V_3^{0h} \cup V_3^0 \rightarrow |||u|||_h = \sqrt{a^h(u, u)}$$

has been performed by P. Lascaux and P. Lesaint [14] using a general error estimate principle (see P.G. Ciarlet [2] for instance):

$$|||u_3 - u_3^h|||_h \leq c \left\{ \inf_{v^h \in V_3^{0h}} ||u - v^h|| + \sup_{w^h \in V_3^{0h}} \frac{|a^h(u, w^h) - l(w^h)|}{|||w^h|||_h} \right\}$$

Let us point out that a restriction on the mesh is necessary which is a real limitation in practical applications. The mesh has to be structured. In the case of triangles, a sufficient condition is that the mesh is generated by three families of parallel straight lines, as it is shown below on Figure III.11.

Figure   III.11

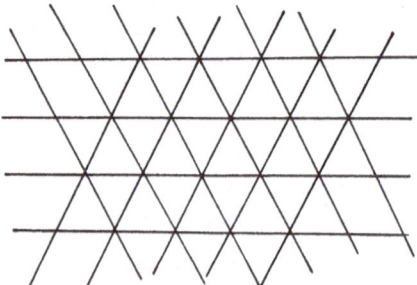

Figure III.12

O.C. Zienkiewicz element	Adini element
TRIANGLES	RECTANGLES
degrees of freedom for the triangle:	*degrees of freedom for the rectangle:*
values at the three summits	*values at the four summits*
2 first-order derivatives at the	*2 values of the first order*
three summits	*derivative at the four summits*
1 value at the gravity center (10 dof version)	
Polynomial space on K is:	*Polynomial space on K is:*

P_3 - {bubble} or P_3
*depending whether the internal
degree of freedom is taken into
account or not*

$$P_K = P_3 \oplus \left\{ x_1\, x_2^3 , x_1^3\, x_2 \right\}$$
$$\dim P_K = 12$$

III.4   The penalty-duality finite element method for the bending plate model

The numerical schemes which are described in this section are based on the Reissner-Mindlin-Naghdi model. As a matter of fact it is necessary to use the so-called modified version, in the preceding sections. Let us recall that the modification concerned only the free edge of the plate. The main consequence of this change is in the boundary layer near the free edge which is destroyed on purpose in the modified model. This enables one to derive sharper error estimates on the continuous solution with respect to the small parameter η. These bounds (basically in the norm $H^2 (\omega)$ for the components θ_α^η and u_3^η) are explained in section III.4.1. Then we focus on the theoretical error analysis for the finite elements based on R.M.N. (**Reissner-Mindlin-**

Naghdi) in section III.4.2. Finally, several examples are discussed in section III.4.4., in which we emphasize the connections with the QUAD 4 element which is so well known by engineers for practical applications.

III.4.1 Stability with respect to the penalty parameter of the R.M.N. solution

Let us describe very shortly the goal of this paragraph. The finite element approximation of a function requires an additional regularity. For instance (cf. paragraph III.1), the error between a function u defined on ω and its interpolate πu, using linear and continuous elements, is such that:

$$\| u - \pi u \|_{1,\omega} \leq c\, h \, |u|_{2,\omega}$$

where h is the mesh size. The difficulty is then to prove that $|u|_{2,\omega}$ is finite. In the R.M.N. model, the solution depends on a small parameter named η. Hence the norm $H^2(\omega)$ of the solution has to be uniformly bounded with respect to η. Otherwise the error bounds would be meaningless.

First of all let us recall the formulation of the R.M.N. model and the main results obtained in Chapter II. The functional spaces used respectively for u_3^η and θ^η are:

$$V_3 = \left\{ v \mid v \in H^1(\omega),\ v = 0 \ \text{ on } \gamma_0 \cup \gamma_1 \right\}$$

and

$$W_t = \left\{ \mu \mid \mu = (\mu_\alpha) \in \left(H^1(\omega)\right)^2 ;\ \mu_\alpha = 0 \ \text{ on } \gamma_0,\ \mu_\alpha\, a_\alpha = 0 \ \text{ on } \gamma_1 \right\},$$

where $\{a_\alpha\}$ denotes the components of the unit tangent to the boundary of ω. Let us recall that γ_0 corresponds to the clamped edge and γ_1 to the simply supported one, the plate being free on γ_2. Then for arbitrary elements θ, μ of the space W_t, we use the bilinear form $k(.,.)$ defined by:

$$k(\theta, \mu) = \frac{2\,E\,\varepsilon^3}{3(1 - v^2)} \int_\omega \left\{ (1 - v)\, \gamma_{\alpha\beta}(\theta)\, \gamma_{\alpha\beta}(\mu) + v\, \gamma_{\lambda\lambda}(\theta)\, \gamma_{\xi\xi}(\mu) \right\}.$$

It has already been mentioned that $k(.,.)$ is W_t-elliptic. Finally the R.M.N. model for bending plates consists in finding an element (θ^η, u_3^η) in the space $W_t \times V_3$ and such that:

$$(\text{III.5}) \quad \begin{cases} \forall\ \mu \in W_t\ ,\ k\left(\theta^\eta, \mu\right) + \dfrac{1}{\eta} \displaystyle\int_\omega \left(\theta_\alpha^\eta + \partial_\alpha u_3^\eta\right)\mu_\alpha = 0\ , \\[4mm] \forall\ v_3 \in V_3\ ,\ \dfrac{1}{\eta} \displaystyle\int_\omega \left(\theta_\alpha^\eta + \partial_\alpha u_3^\eta\right)\partial_\alpha v_3 = \displaystyle\int_\omega f_3\, v_3\ . \end{cases}$$

It has been proved in Chapter II that (III.5) has a unique solution. Furthermore we proved in Theorem II.2 that for small enough η, there exists a constant c independent of η and such that (if we consider the non-modified R.M.N. model with a free edge):

$$(\text{III.6}) \quad \begin{cases} \displaystyle\sum_{\alpha=1,2} \left\|\theta_\alpha^\eta + \partial_\alpha u_3^\eta\right\|_{0,\omega} \le c\,\sqrt{\eta}\ , \\[4mm] \left\|\theta^\eta\right\|_{W_t} \le c\ , \\[4mm] \left\|u_3^\eta\right\|_{1,\omega} \le c\ . \end{cases}$$

which could be easily obtained from (III.5) by setting $\mu = \theta^\eta$ and $v_3 = u_3^\eta$ and using the W_t-coerciveness of the bilinear form $k(.,.)$.

The first equation (III.5) can be locally interpreted by:

$$-\partial_\beta m_{\alpha\beta}^\eta = -\frac{1}{\eta}\left(\theta_\alpha^\eta + \partial_\alpha u_3^\eta\right)$$

where :

$$m_{\alpha\beta}^\eta = \frac{2\,E\,\epsilon^3}{3\left(1 - v^2\right)}\left\{(1 - v)\,\gamma_{\alpha\beta}\left(\theta^\eta\right) + v\,\gamma_{\lambda\lambda}\left(\theta^\eta\right)\delta_{\alpha\beta}\right\}.$$

This is a classical two-dimensional elasticity operator. Obviously one has to add the boundary conditions depending on the part of the boundary which is concerned. Let us assume (this is the standard approach in finite element methods) that the two-dimensional elasticity operator equipped with the boundary conditions is smooth. In other words, we assume that $\theta_\alpha^\eta \in H^2(\omega)$ and that there exists a constant c such that:

$$(\text{III.7}) \quad \sum_{\alpha=1,2} \left\|\theta_\alpha^\eta\right\|_{2,\omega} \le \frac{c}{\eta}\left\{\sum_{\alpha=1,2} \left\|\theta_\alpha^\eta + \partial_\alpha u_3^\eta\right\|_{0,\omega}\right\},$$

such regularity results can obviously be discussed separately. They are purely mathematical tricks. The main restrictions in this case are:

– *the boundaries γ_0, γ_1 and γ_2, have to be smooth (C^1 is sufficient);*

– at the connection between the various components of γ_0, γ_1 and γ_2, special continuity conditions have to be satisfied. For instance, γ_0 and γ_1 should intersect at a right angle. The simplest case for which they are satisfied, corresponds to disconnected boundary components.

For a complete analysis of this problem, we refer to P. Grisvard [17]. Let us then point out that a combination of (III.6) and (III.7) leads to the following estimate:

(III.8)
$$\sum_{\alpha=1,2} \left\| \theta_\alpha^\eta \right\|_{2,\omega} \leq \frac{c}{\sqrt{\eta}} \ .$$

Furthermore, even if it is not straightforward, it seems quite impossible to improve such an upper bound. The fact is that the R.M.N. model presents a boundary layer effect which involves terms like $\xi\, e^{-\frac{\xi}{\sqrt{\eta}}}$, ξ being the normal coordinate along the boundary of ω. This boundary layer is mainly localized in the vicinity of the free edge (for details concerning the construction of this boundary layer, we refer to J.L. Lions [18] for general formulation and to Ph. Destuynder [19] for plates). From a descriptive point of view, the reader can find interesting indications in K.O. Friedrichs - R. F. Dressler [20] and in A.L. Gol'denveizer [21].

Let us also note that the norm of terms like $\xi\, e^{-\frac{\xi}{\sqrt{\eta}}}$ in $H^2(\omega)$ norm is equivalent to $\dfrac{1}{\eta^{1/4}}$ and thus

cannot be bounded with respect to η, when η tends to zero. Let us now come back to the P_1-Q_1 Lagrange interpolate of θ_α^η . The classical error estimates of Ciarlet [2] and Raviart-Thomas [3], recalled in section III.1, leads to :

$$\left\| \theta_\alpha^\eta - \pi\, \theta_\alpha^\eta \right\|_{1,\omega} \leq c\, \frac{h}{\sqrt{\eta}}$$

which can be useful if $h \ll \sqrt{\eta}$. Unfortunately, η is chosen in order to have a physical meaning of the penalty model. More precisely, one has usually:

$$\eta \approx \varepsilon^2 ,$$

compared to the stiffness of the term $k(.,.)$, which is homogeneous to $D = \dfrac{2\,E\,\varepsilon^3}{3\,(1-\nu^2)}$, i.e.

$\eta \approx \dfrac{\varepsilon^2}{D} \approx \dfrac{(1+\nu)}{2\,E\,\varepsilon}$ which is the transverse compliance. Hence:

$$\frac{h}{\sqrt{\eta}} \approx \frac{h}{\varepsilon} ,$$

For a thin plate (for instance $\varepsilon = 10^{-2}\,L$, L being the diameter of ω), one should almost have:

$$h \approx \varepsilon^{3/2} \ ,$$

in order to keep the order of the error estimate between the three-dimensional solution and the solution to the plate model, which is : $o(\sqrt{\varepsilon})$ near the edges (see Ph. Destuynder [9]). Obviously such a condition is too restrictive and thus the standard penalty (or R.M.N.) model cannot be used, at least if there is a free edge.

Let us now discuss the modified penalty model. It is clear that both are identical if the free edge component of the boundary of ω is empty. It has been proved in section II.3.2.3 (Theorem II.3) that if $(\theta^{\eta}, u_3^{\eta})$ is solution to this *modified model*, then:

(III.9) $$\sum_{\alpha = 1, 2} \left\| \theta_{\alpha}^{\eta} + \partial_{\alpha} u_3^{\eta} \right\|_{0, \omega} \leq c \, \eta$$

(see (II.55). Hence, the regularity assumption for the elasticity operator leads to:

(III.10) $$\sum_{\alpha = 1, 2} \left\| \theta_{\alpha}^{\eta} \right\|_{2, \omega} \leq \frac{c}{\eta} \left\{ \sum_{\alpha = 1, 2} \left\| \theta_{\alpha}^{\eta} + \partial_{\alpha} u_3^{\eta} \right\|_{0, \omega} \right\} = c \sum_{\alpha = 1, 2} \left\| q_{\alpha}^{\eta} \right\|_{0, \omega} \leq c_1 \, ,$$

where c_1 is a constant which does not depend on the small parameter η. Hence, using P_1 - Q_1 - Lagrange elements, the interpolation results lead to :

(III.11) $$\left\| \theta_{\alpha}^{\eta} - \pi \, \theta_{\alpha}^{\eta} \right\|_{1, \omega} \leq c_2 \, h$$

Thus it clearly appears that the modified version of the penalty model for bending plates is quite necessary for obtaining a uniformly stable finite element scheme with respect to η. This remark is very important. But it should be also noticed that the estimate (III.10) is also valid for the non modified R.M.N. system as soon as γ_2 is empty.

Nevertheless, in order to avoid technical difficulties (concentrated in the writing but not in the basic concepts), we focus in the following (section III.4), to the case where γ_2 is empty. The principles are the same in the general case but with an additional complexity due to the terms on γ_2. We leave it to the reader.

With another respect, the u_3^{η} component of R.M.N. model is solution to (see (III.5)):

$$\begin{cases} - \Delta u_3^{\eta} = \eta \, f_3 + \partial_{\alpha} \theta_{\alpha}^{\eta} \ \text{on} \ \omega \ , \\ u_3^{\eta} = 0 \ \text{on} \ \gamma_0 \cup \gamma_1 = \partial \omega \ ! \end{cases}$$

γ_2 is assumed to be empty from now on up to the end of section III.4. Assuming the regularity of Laplace operator on ω with homogeneous Dirichlet boundary conditions enables one to obtain the estimate:

$$\text{(III.12)} \qquad \left\|u_3^\eta\right\|_{2,\omega} \le c \left\{\eta\|f_3\|_{0,\omega} + \left\|\theta^\eta\right\|_{W_t}\right\} \le c$$

where c is a constant independent of η, as soon as η is small enough. For additional justifications on this regularity assumption, we refer again to P. Grisvard [17], even if this situation (Poisson equation) is very classical.

Let us now go to technical results in interpolation theory which are helpful in the following.

III.4.1.1   *Approximation of the transverse shear stress using discontinuous element*

Let us first recall that, in the R.M.N. model, the resultant transverse shear stress is given by the following formula :

$$q_\alpha^\eta = \frac{1}{\eta}\left(\theta_\alpha^\eta + \partial_\alpha u_3^\eta\right) \qquad \alpha = 1, 2$$

and from (III.9) one has:

$$\text{(III.13)} \qquad \left\|q_\alpha^\eta\right\|_{0,\omega} \le c$$

where c is an η-independent constant. Let us point out here again that the boundary corresponding to the free edge has been assumed to be empty in order to simplify the notations. Furthermore, it would not be acceptable to assume that the norm $H^1(\omega)$ of the element q_α^η is bounded uniformly with respect to η (see Chapter II). Thus the error between q_α^η and an approximate field should be analyzed in the restricted framework of (III.13). The functional space in which the asymptotic behaviour of the R.M.N. solution has been studied when η tends to zero is such that:

$$q_\alpha^\eta \in H^{-1}(\omega) \quad \text{and} \quad \operatorname{div} q^\eta = \partial_\alpha q_\alpha^\eta \in H^{-1}(\omega) .$$

Therefore the error analysis in the approximation of q_α^η has to be considered in this framework. The main point in the choice done in this section III.4 is to build a discontinuous approximation of q_α^η which enables one to eliminate locally this term in the effective solution of the R.M.N. model. The difficulty that we meet is to have elements rich enough in order to furnish an approximation of q_α^η but also of $\operatorname{div} q^\eta$ and such that the discrete Brezzi conditions [22] that are presented at (III.22) and (III.23) should be satisfied with constants which are independent of both η (the penalty parameter in R.M.N. model) and h (the mesh size in the finite element approximation of functions over the open set ω which is the medium surface of the plate). Let us recall that an element $q = (q_\alpha)$ lying in the space $\left(L^2(\omega)\right)^2$ can be split into the sum of a gradient and a rotational. More precisely, if we additionally assume that div q $\in$ L$^2(\omega)$ (which is a weaker assumption than $q_\alpha \in$ H$^1(\omega)$!) we can write (see Lemma II.3):

$$q = \text{grad } \varphi + \text{rot } \psi$$

where both φ and ψ lie in the space H$^1(\omega)$. Furthermore, we can prescribe $\varphi = D_k$, on the K-1 first connected components of the boundary of ω and zero on the last one denoted: $\partial\omega_K$; then ψ is uniquely defined up to a constant. If div q $\in$ L$^2(\omega)$, **and if for sake of simplicity** $\partial\omega$ **is simply connected**, then φ is solution of:

$$\begin{cases} - \Delta \varphi = - \text{div q} \in L^2(\omega), \text{ on } \omega, \\ \quad \varphi = 0 \text{ on } \partial\omega \end{cases}$$

Assuming the regularity in the Dirichlet problem (see P. Grisvard [17]), one can assume that $\varphi \in$ H$_0^1(\omega) \cap$ H$^2(\omega)$ and then there exists a constant c such that:

$$\|\varphi\|_{2,\omega} \leq c \|\text{div q}\|_{0,\omega}$$

(obviously c does not depend on q). As we mentioned earlier, this assumption is satisfied as soon as the boundary $\partial\omega$ of ω is for instance C^1. Let us now go to the definition of an approximation space for the transverse shear stress q$^\eta$. We set:

(III.14)
$$H_t^{ho} = \{q \mid q = (q_\alpha), q_\alpha \in L^2(\omega), \forall K \in F^h \; q_{|K} = \text{grad } \varphi_K \text{ with } \varphi_K \in P_1, \\ \forall K \in Q^h, q_{|K} = \text{grad } \varphi_K + \text{rot } \psi_K \\ \text{where } \varphi_K \text{ and } \psi_K \text{ are both functions of the space } Q_1\}$$

Let us mention here that for sake of clarity, we make use of the following convention:

• F^h is the collection of triangles in the mesh T^h,
• Q^h is the collection of quadrilaterals in the mesh T^h.

An important point (may be the most important) is that both φ_K and ψ_K are discontinuous from one element to the other. This implies also that derivatives of q^h contain distributions along the sides of the elements of T^h. Thus it cannot be assumed that: div q$^h \in$ L$^2(\omega)$ if q$^h \in$ H$_t^{ho}$.

III.4.1.1.1 *Construction of a projection operator from* H$_t$ *into* H$_t^{ho}$

For sake of brevity in the notations we assume that the boundary $\partial\omega$ is simply connected; see Chapter II.

As a matter of fact, an approximation of an element q of the space H_t is defined from the decomposition of q into the sum of a gradient and a rotational. But it is not possible to use the classical interpolation results in Sobolev spaces as in Ciarlet [2] and Raviart-Thomas [3]. This is due to the fact that such interpolation methods require an additional regularity on q ($q_\alpha \in H^2(\omega)$ at least !) which is unfortunately not satisfied in our case (uniformly with respect to the penalty parameter η). **The only one we can use is** :

$$\boxed{q_\alpha \in L^2 \quad \text{and} \quad \text{div } q \in L^2}$$ (see section *III.4.1.1*).

Approximation of φ

First of all, let us introduce a finite element space based on a continuous approximation of functions which are piecewise linear. Thus we set:

(III.15) $V_0^h = \left\{ v \mid v \in C^0(\overline{\omega}), \ \forall K \in F^h, \ v_{|K} \in P_1, \ \forall K \in Q^h \ v_{|K} \in Q_1, \ v = 0 \text{ on } \partial \omega \right\}$.

Let us then define the element φ^h in the space V_0^h such that:

$$\forall \ v \in V_0^h \quad \int_\omega \left(\text{grad} \left(\varphi - \varphi^h \right), \ \text{grad } v \right) = 0 \ ,$$

where φ is an element of the space $H_0^1(\omega)$. This approximation of φ is very classical (projection from $H_0^1(\omega)$ onto V_0^h), and the scalar product $(u , v)_{1,\omega} = \int_\omega (\text{grad } u , \text{grad } v)$ is clearly (from Poincaré inequality) equivalent to the one of $H^1(\omega)$ on the subspace $H_0^1(\omega)$. If φ is also an element of the space $H^2(\omega)$ which has been assumed earlier because it is the gradient component of q and div q is supposed to be in $L^2(\omega)$, one has the following error estimate (see again Ciarlet [2] or Raviart-Thomas [3]):

$$\left\| \varphi - \varphi^h \right\|_{1,\omega} \leq c \, h \, |\varphi|_{2,\omega}$$

or else, because of the regularity assumption of the solution to the Poisson equation on ω (see section III.4.1.1):

$$\left\| \varphi - \varphi^h \right\|_{1,\omega} \leq c \, h \, \| \text{div } q \|_{0,\omega} \ .$$

From this result, it can be deduced straightforwardly that for the two components of the gradient one has:

$$\left\| \left(\text{grad } \varphi - \text{grad } \varphi^h \right)_\alpha \right\|_{-1,\omega} \leq c \, h \, \| \text{div } q \|_{0,\omega}$$

but also that:

$$\left\| \text{div} \left(\text{grad } \varphi \right) - \text{div} \left(\text{grad } \varphi^h \right) \right\|_{-1,\omega} = \sup_{z \in H_0^1(\omega)} \frac{\displaystyle\int_\omega \left(\text{grad} \left(\varphi - \varphi^h \right), \ \text{grad } z \right)}{\| z \|_{1,\omega}} \leq c \, h \, \| \text{div } q \|_{0,\omega} \ .$$

Approximation of ψ

Let us now introduce an approximation of the rotational component of q which is an element of H_t. We recall that (see Lemma II.3):

$$q = \text{grad } \varphi + \text{rot } \psi$$

where $\psi \in H^1(\omega) \cap L_0^2(\omega)$ because $q_\alpha \in L^2(\omega)$ and $\varphi \in H_0^1(\omega)$. We recall again that:

$$L_0^2(\omega) = \left\{ v \, | \, v \ \in L^2(\omega), \ \int_\omega v = 0 \right\} \ .$$

We define the approximation space by :

$$W_0^h = \left\{ z \, | \, z \in C^0(\overline{\omega}); \ \forall \, K \in F^h, \ z_{|K} \in P_1, \ \forall \, K \in Q^h, z_{|K} \in Q_1 ; \ \int_\omega z = 0 \right\} \ .$$

The first step of our analysis consists in noticing that the bilinear form:

$$(z, v) \in H^1(\omega) \cap L_0^2(\omega) \ \rightarrow \ \int_\omega (\text{rot } z, \ \text{rot } z)$$

is coercive on the space $H^1(\omega) \cap L_0^2(\omega)$ and therefore define an equivalent scalar product on this space. This is a classical result left to the reader. Thus there exists a positive constant – say c – such that:

$$\forall \ z \in H^1(\omega) \cap L_0^2(\omega) \qquad \int_\omega |\text{rot } z|^2 \geq c \, \| \text{rot } z \|_{1,\omega}^2$$

Then we define ψ^h element of the space W_0^h and such that:

$$\forall \ v \in W_0^h, \ \int_\omega \left(\mathrm{rot} \ \psi^h - \mathrm{rot} \ \psi , \ \mathrm{rot} \ v \right) = 0 \ .$$

Since ψ is not necessarily in the space $H^2(\omega)$, it is not possible to apply general error estimates as we did for $\varphi - \varphi^h$. A new trick is required. As a matter of fact, we are going to prove that there exists a constant – say c_1 – such that:

$$\left\| \psi - \psi^h \right\|_{0,\omega} \le c_1 \, h \, \left\| \psi \right\|_{1,\omega}$$

and because :

$$\mathrm{rot} \ \psi = q - \mathrm{grad} \ \varphi$$

we shall deduce that: (φ being defined only with respect to q in the case of a simply connected boundary; concerning the general case see chapter II. section 4.8)

$$\left\| \psi - \psi^h \right\|_{0,\omega} \le c_2 \, h \, \left\| q \right\|_{H_t} \ , \ \left(\left\| q \right\|_{H_t} = \sqrt{ \sum_{\alpha=1,2} \left\| q_\alpha \right\|_{0,\omega}^2 } \right)$$

($\varphi \in H_0^1(\omega)$ is solution to: $- \Delta \varphi = - \mathrm{div} \ q$, therefore $\| \varphi \|_{1,\omega} \le c_2 \, \| q \|_{H_t}$).

Let us first notice that for any functions λ and λ^h one has:

$$\left\| \lambda - \lambda^h \right\|_{0,\omega} = \sup_{t \in L^2(\omega)} \frac{ \int_\omega \left(\lambda - \lambda^h \right) t }{ \| t \|_{0,\omega} } \ .$$

Then we introduce the element $z \in H^1(\omega) \cap L_0^2(\omega)$ associated to t and such that:

$$\forall \ v \in H^1(\omega) \cap L_0^2(\omega), \ \int_\omega \left(\mathrm{rot} \ z , \ \mathrm{rot} \ v \right) = \int_\omega t \, v$$

and its W_0^h -approximate, z^h, defined by:

$$z^h \in W_0^h, \ \forall \ v \in W_0^h, \ \int_\omega \left(\mathrm{rot} \ z^h , \ \mathrm{rot} \ v \right) = \int_\omega t \, v \ .$$

Thus:

$$\|\psi - \psi^h\|_{0,\omega} = \sup_{t \in L^2(\omega)} \frac{\int_\omega \left(\mathrm{rot}\, z\,,\, \mathrm{rot}\left(\psi - \psi^h\right)\right)}{\|t\|_{0,\omega}}$$

$$= \sup_{t \in L^2(\omega)} \frac{\int_\omega \left(\mathrm{rot}\left(z - z^h\right),\, \mathrm{rot}\left(\psi - \psi^h\right)\right)}{\|t\|_{0,\omega}}$$

$$\leq \sup_{t \in L^2(\omega)} \left(\frac{\|z - z^h\|_{1,\omega}}{\|t\|_{0,\omega}}\right) \|\psi - \psi^h\|_{1,\omega}.$$

But on the one hand, the definition of ψ^h enables one to derive immediately the estimate:

$$\|\psi^h\|_{1,\omega} \leq c \|\psi\|_{1,\omega}$$

and on the other hand, assuming the regularity of the Neuman problem on ω :

$$z \in H^1(\omega) \cap L^2_0(\omega),\ -\Delta z = t_0 \in L^2_0(\omega),\ \frac{\partial z}{\partial b} = 0 \text{ on } \partial\omega,\ \Rightarrow\ z \in H^2(\omega),$$

we can write (see again Ciarlet [2] or Raviart-Thomas [3]):

$$\|z - z^h\|_{1,\omega} \leq c_3\, h |z|_{2,\omega} \leq c_4\, h\, \|t\|_{0,\omega}$$

(details are left to the reader, setting: $t_0 = t - \frac{1}{|\omega|} \int_\omega t$).

Finally we obtain :

$$\|\psi - \psi^h\|_{0,\omega} \leq c_5\, h\, \|\psi\|_{1,\omega}$$

which is the claimed estimate and where ψ^h is defined by:

$$\begin{cases} \psi^h \in W^h_0, \\ \forall\ v \in W^h_0,\ \int_\omega \left(\mathrm{rot}\left(\psi - \psi^h\right),\, \mathrm{rot}\, v\right) = 0\ . \end{cases}$$

We are now able to state the basic result:

Theorem III.4

The approximate of an element q lying in the space H_t and such that: div q *is in* $L^2(\omega)$, *by an element of the space* $H_t^{h\,0}$ *is defined by:*

$$P^h q = \text{grad } \varphi^h + \text{rot } \psi^h$$

where   q = grad φ + rot ψ and φ^h, ψ^h are defined previously. Then one has the following estimates for $\alpha = 1, 2$ (we use the continuity of the derivative from $L^2(\omega)$ into $H^{-1}(\omega)$ and the fact that: div(rot(.)) =0):

$$(\text{III}.16) \quad \begin{cases} \left\| \left(P^h q - q \right)_\alpha \right\|_{-1,\omega} \le c\,h \, \|q\|_{H_t} \\ \left\| \text{div} \left(P^h q - q \right) \right\|_{-1,\omega} \le c\,h \, \|\text{div } q\|_{0,\omega} \end{cases}$$

∎

The proof of III.16 is just a compilation results obtained in this sub-section III.4.1.1.1.and those obtained previously.

∎

Remark III.4

It is important to notice that the components of the element q are not necessarily in the space $H^1(\omega)$.

∎

III.4.1.2   Approximation of the rotation θ^η solution to the R.M.N. model

The rotation field $\theta^\eta = \left(\theta_\alpha^\eta \right)$ lies in the space W_t, the definition of which has been recalled at the beginning of section III.4. Basically the components θ_α^η are in the space $H^1(\omega)$ and satisfy some boundary conditions on $\partial\omega$. The approximation space of W_t which is introduced hereafter is denoted by $W_t^{h\,0}$ and has two or more internal degree of freedom for each element of the mesh T^h (two for triangles and four for quadrilaterals).

a) TRIANGLES

Let us introduce the functions connected to these internal degrees of freedom. First of all we define the so-called Ramses function equal to one at the center of gravity and zero on the boundary of the element. Furthermore it is piecewise linear on each of the sub-triangles or sub-quadrilaterals as shown on Figure III.13.

b) QUADRILATERALS

For quadrilaterals another internal function is used also associated to two internal degrees of freedom. It has to be linearly independent of the Ramses function. The simplest idea is certainly to consider the so-called Bubble function on the element K of T^h. It is defined on the reference element (which is a square!) as a second-degree polynomial equal to one at the center of gravity of the element and vanishing on the boundary of K. We denote it by B_K.

Figure III.13

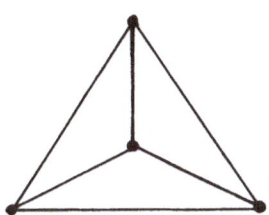 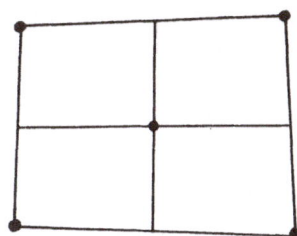

Then the internal degrees of freedom of the elements of T^h are defined as the coefficients A_1 and A_2 for triangles or A_1, A_2, A_3 and A_4 for quadrilaterals, such that the following local (i.e. with support included in a single element K) functions are added to the classical P_1 or Q_1 finite element space (for each component θ_α^η):

•   for triangles:
     $R_{KT} = \{A_1(R_K, 0), A_2(0, R_K)\}$

•   for quadrilaterals:
     $B_{KQ} = \{A_1(B_K, 0), A_2(0, B_K)\}$
     $R_{KQ} = \{A_3(-x_1, x_2) R_K, A_4(x_2, x_1) R_K\}$

(let us point out that A_i, i = 1, 2, 3, 4 are arbitrary constants and x_α; α = 1, 2 are the cartesian coordinates in the reference frame to which $\hat{K}$ is referred; $\hat{K}$ being the reference element). In order to avoid additional difficulty, we suppose that the quadrilaterals are rectangles. This permits to confuse the coordinates x_α of the reference set of axes to which $\hat{K}$ is referred, with those to which ω is referred. In fact, the mapping from $\hat{K}$ to K is translation-rotation, combined with an affinity along the coordinate axes.

Remark III.5
Arbitrary quadrilaterals could also be used with just an additional complexity in the notations (but not in the mathematics). Then the shape functions (internal or not) are polynomials with respect to the coordinates of the reference element $\hat{K}$. But on the current element they are much more complicated. The reader is referred to section III.1 for further details. ∎

The approximation space for θ^η can now be defined by:

$$(\text{III.17}) \quad W_t^{h\,0} = \left\{ \mu \,|\, \mu = (\mu_\alpha) \in W_t \,;\, \forall\, K \in F^h \; \mu_{\alpha|K} \in P_1 \,;\, \forall\, K \in Q^h \,,\, \mu_{\alpha|K} \in Q_1 \right\}$$

$$\oplus \quad \prod_{K \,\in\, F^h} R_{KT} \quad \oplus \quad \prod_{K \,\in\, Q^h} B_{KQ} \quad \oplus \quad \prod_{K \,\in\, Q^h} R_{KQ}$$

(F^h denotes the triangles of the mesh T^h and Q^h denotes the quadrilaterals of T^h). Our goal is then to define an interpolate – say $\pi\theta$ – of an element θ of the space W_t and such that:

- $\pi\,\theta \in W_t^{h\,0}$

- $\pi\,\theta$ is very close to θ when h tends to zero.

We proceed as follows. The values of $\pi\,\theta$ at the vertices of the mesh are those of θ (i.e. the values at the summit of each triangle or quadrilateral):

$$\forall\, a_i \text{ vertex of } T^h\!: \; \pi\,\theta\,(a_i) = \theta\,(a_i) \; .$$

Concerning the internal degrees of freedom, several strategies are possible as far as the only goal is to obtain a consistent approximation of θ as h tends to zero. They can even be chosen equal to zero. But for the checking of the so-called L.B.B. (Ladyzenskaïa - Babuska - Brezzi) discrete condition in the following, it is very convenient to use another definition. Thus we define these internal degrees of freedom by the conditions III.18 which are explicited hereafter.

(III.18)

- if $K \in F^h$ (K is a triangle)

$$\forall \ \psi \in P_1 \ (on \ K), \ \int_K \left(\pi \ \theta - \theta , \ rot \ \psi\right) = 0,$$

$$\forall \ \varphi \in P_1 \ (on \ K), \ \int_K \left(\pi \ \theta - \theta , \ grad \ \varphi\right) = 0;$$

- if $K \in Q^h$ (K is a quadrangle)

$$\forall \ \psi \in Q_1 \ (on \ K), \ \int_K \left(\pi \ \theta - \theta , \ rot \ \psi\right) = 0,$$

$$\forall \ \varphi \in Q_1 \ (on \ K), \ \int_K \left(\pi \ \theta - \theta , \ grad \ \varphi\right) = 0.$$

When K is a triangle, the relations (III.18) consist in setting the mean value of $\pi \ \theta - \theta$ on each element to zero. As a matter of fact, let us emphasize that the two conditions (one with φ and the other with ψ) are equivalent. When K is a quadrilateral, the relations (III.18) are slightly more difficult to discuss. Let us also point out that for arbitrary quadrilaterals, it is necessary to formulate (III.18) on the reference element $\widehat{K}$. Our purpose is now to prove that (III.18) enables one to define the two or four internal degrees of freedom on K.

First of all we clearly have:

$$\begin{cases} \int_K \left(\pi \ \theta - \theta , \ rot \ \psi\right) = - \int_K \left(\pi \ \theta_1 - \theta_1\right) \partial_2 \ \psi + \int_K \left(\pi \ \theta_2 - \theta_2\right) \partial_1 \ \psi, \\ \int_K \left(\pi \ \theta - \theta , \ grad \ \varphi\right) = \int_K \left(\pi \ \theta_1 - \theta_1\right) \partial_1 \ \varphi + \int_K \left(\pi \ \theta_2 - \theta_2\right) \partial_2 \ \varphi. \end{cases}$$

and if set for quadrilaterals (C and G are zero for triangles):

$$\begin{cases} \psi = A \, x_1 + B \, x_2 + C \, x_1 \, x_2 + D \\ \varphi = E \, x_1 + F \, x_2 + G \, x_1 \, x_2 + H \end{cases}$$

we obtain:

$$\bullet \int_K \left(\pi\,\theta - \theta \,,\ \text{rot } \psi \right) = A \int_K \left(\pi\,\theta_2 - \theta_2 \right) - B \int_K \left(\pi\,\theta_1 - \theta_1 \right)$$

$$+ \ C \int_K \left[x_2 \left(\pi\,\theta_2 - \theta_2 \right) - x_1 \left(\pi\,\theta_1 - \theta_1 \right) \right]$$

$$\bullet \int_K \left(\pi\,\theta - \theta \,,\ \text{grad } \varphi \right) = E \int_K \left(\pi\,\theta_1 - \theta_1 \right) + F \int_K \left(\pi\,\theta_2 - \theta_2 \right)$$

$$+ \ G \int_K \left[x_2 \left(\pi\,\theta_1 - \theta_1 \right) + x_1 \left(\pi\,\theta_2 - \theta_2 \right) \right]$$

and thus the relations (III.18) are equivalent to:

$$\bullet \int_K \left(\pi\,\theta_1 - \theta_1 \right) = \int_K \left(\pi\,\theta_2 - \theta_2 \right) = 0$$

$$\bullet \int_K x_2 \left(\pi\,\theta_2 - \theta_2 \right) - \int_K x_1 \left(\pi\,\theta_1 - \theta_1 \right) = 0 \qquad \cdot$$

$$\bullet \int_K x_2 \left(\pi\,\theta_1 - \theta_1 \right) + \int_K x_1 \left(\pi\,\theta_2 - \theta_2 \right) = 0$$

But on the quadrilateral K (assumed to be a rectangle for sake of simplicity), $\pi\theta$ is the sum of the classical Q_1 Lagrange interpolate – say $\pi^1\theta$ – and a function – say $R\,(\theta)$ – which is associated to the four internal degrees of freedom of the element. Thus let us set on K:

$$\begin{cases} \pi\,\theta_1 = \pi^1\,\theta_1 + A_1\,B_K - A_3\,x_1\,R_K + A_4\,x_2\,R_K \,, \\ \pi\,\theta_2 = \pi^1\,\theta_2 + A_2\,B_K + A_3\,x_2\,R_K + A_4\,x_1\,R_K \,, \end{cases}$$

where B_K (respectively R_K) is the so-called "Bubble function" (respectively the "Ramses function"). Let us notice that a rectangle is balanced with respect to its center of gravity. In other words, K is symmetrical with respect to this point. Thus:

$$\int_K x_1\,R_K = \int_K x_2\,R_K = \int_K x_1\,B_K = \int_K x_2\,B_K = 0 \ .$$

Finally the relations (III.18) are equivalent to

$$A_1 \int_K B_K \ = \ C_1 \ ,$$

$$A_2 \int_K B_K \ = \ C_2 \ ,$$

$$A_3 \int_K \left(x_1^2 + x_2^2\right) R_K \ = \ C_3 \ ,$$

$$A_4 \int_K \left(x_1^2 + x_2^2\right) R_K \ = \ C_4 \ ,$$

where C_1, C_2, C_3 and C_4 are four constants on K depending on $\theta - \pi^1\theta$ ($\pi^1\theta$ is the Q_1 Lagrange interpolate of θ). Furthermore it appears clearly from the previous relations that the term $R(\theta)$ (associated to the four internal degrees of freedom and defined by: $R(\theta) = \pi\theta - \pi^1\theta$) is continuously defined with respect to $\theta - \pi^1\theta$ as an element of the space $L^2(\omega)$ for instance.

This enables one to derive an error bound between θ and $\pi\theta$. Noticing that:

$$\|\theta - \pi\theta\|_{H_t} \leq \|\theta - \pi^1\theta\|_{H_t} + \|R(\theta)\|_{H_t}$$

we deduce that (for a constant c which is h-independent):

(III.19) $\|\theta - \pi\theta\|_{H_t} \leq c\|\theta - \pi^1\theta\|_{H_t} \leq c_1 h^2 \left(\sum_{\alpha=1,2} |\theta_\alpha|_{2,\omega} \right)$

and in a similar manner:

$$\|\theta - \pi\theta\|_{W_t} \leq \|\theta - \pi^1\theta\|_{W_t} + \|R(\theta)\|_{W_t}$$

which leads to (using the inverse inequality $\|R(\theta)\|_{W_t} \leq \dfrac{c}{h}\|R(\theta)\|_{H_t}$; see Ciarlet [2]):

(III.20) $\|\theta - \pi\theta\|_{W_t} \leq ch \left(\sum_{\alpha=1,2} |\theta_\alpha|_{2,\omega} \right) \ .$

Once more the constants c appearing in the preceding inequalities are h-independent and furthermore are distinct one from the other.

Remark III.6 *"Very important for the following"*
Let us consider a couple of elements (θ , u_3) which belongs to the space $W_t \times V_3$ (see paragraph III.4.1 for the definition of the space V_3). Then the P_1-Q_1 Lagrange interpolate of u_3 is denoted

by $\pi^1 u_3$. It belongs to the space V_0^h mentioned previously in section III.4.1.1. Furthermore, one has the classical error estimate (see Ciarlet [2] or Raviart-Thomas [3]):

$$\| u_3 - \pi^1 u_3 \|_{1,\omega} \leq c h \, |u_3|_{2,\omega}$$

where c (as usual) is a constant which is independent of h and of u_3. There is now a new way to define the interpolate of θ which will be very useful in the following. But it is dependent on u_3. This makes sense when the penalty parameter η is very small because at the limit (when $\eta = 0$), one has:

$$\theta + \text{grad } u_3 = 0.$$

Let us set for this new interpolate of θ:

$$\pi^3 \theta = \pi^1 \theta + R(\theta, u_3)$$

where $\pi^1 \theta$ is the classical P_1-Q_1-Lagrange interpolate as we did previously. The function $R(\theta, u_3)$ represents the contribution of the internal degrees of freedom. They were chosen such that:

$$\forall \, p \in H_t^{h0}, \quad \int_\omega p_\alpha (\theta - \pi \theta)_\alpha = 0.$$

But in this remark, we choose the new following relations for defining $R(\theta, u_3)$:

$$\forall \, p \in H_t^{h0}, \quad \int_\omega p_\alpha (\theta_\alpha - \pi^3 \theta)_\alpha = - \int_\omega p_\alpha \partial_\alpha (u_3 - \pi^1 u_3).$$

Then one has with similar justifications as for (III.20):

$$\| \theta - \pi^3 \theta \|_{W_t} \leq c h \left\{ \sum_{\alpha=1,2} |\theta_\alpha|_{2,\omega} \right\} + \| u_3 - \pi^1 u_3 \|_{1,\omega}$$

or finally, with another constant c:

(III.21) $$\| \theta - \pi^3 \theta \|_{W_t} \leq c h \left\{ \sum_{\alpha=1,2} |\theta_\alpha|_{2,\omega} + |u_3|_{2,\omega} \right\}.$$

Obviously, here again, c is independent of h, θ and u_3. This new interpolate $\pi^3 \theta$ which is linked to u_3 (this is why we used the superscript 3) is the one that we use in section III.4.3. for the final error estimate between the solutions to the R.M.N. model and its approximation by a "local transverse shear stress finite-element method".   ∎

III.4.1.3   *Abstract error estimates in the finite element approximation of the R.M.N. model*

Let us first recall briefly that we have decided in the current section (III.4) to ignore the problems raised by the boundary layer near the free edges by assuming that "γ_2 is empty". Therefore we deal with the penalty model defined in (III.5) which exactly coincides in this case with the modified penalty version. But the extension of the analysis given hereafter to the general situation is quite obvious and is left to the reader. The only point that we wish to point out is the necessity to use the modified (in fact, upgraded) penalty model in practical applications where a free boundary is involved. Nevertheless one could say that far from the boundary layer the two solutions (penalty model or modified one) are very close.

The first step of our analysis consists in formulating the compatibility conditions between the approximation spaces. The first one states that u_3 can be controlled from a duality term, and the second one ensures that the dual variables can also be controlled by the primal ones. They are explicited as follows:

$$\text{(III.22)} \quad \exists\ c_1 > 0,\ \forall\ v_3 \in V_3^h,\ \sup_{p \in H_t^h} \frac{\displaystyle\int_\omega p_\alpha\, \partial_\alpha\, v_3}{\|p\|_{H_t}} \geq c_1 \|v_3\|_{1\,,\,\omega}$$

$$\text{(III.23)} \quad \exists\ c_2 > 0,\ \forall\ p \in H_t^h,\ \sup_{(\mu\,,\,v_3) \in W_t^h \times V_3^h} \frac{\displaystyle\int_\omega p_\alpha\,(\mu_\alpha + \partial_\alpha\, v_3)}{\|\mu\|_{W_t} + \|v_3\|_{1\,,\,\omega}} \geq c_2\, h\, \|p\|_{H_t}$$

where c_1 and c_2 are two real constants which are independent of the discretization parameter h, which has the meaning of a mesh size. Furthermore, we define the following norm:

$$\forall\ p \in H_t^h,\ \|p\|_{M_t^h} = h\,\|p\|_{H_t}\ .$$

As a matter of fact, this norm – very convenient for the proofs – is h-dependent.

Finally, let us introduce three finite dimensional subspaces of W_t, V_3, H_t, denoted by: W_t^h, V_3^h, H_t^h.

Let us now introduce the approximate model by :

(III.24)
$$
\begin{cases}
\text{find}\left(\theta^{\eta\,h},\, u_3^{\eta\,h},\, q^{\eta h}\right) \in W_t^h \times V_3^h \times H_t^h \text{ such that :} \\[2mm]
\forall\, \mu \in W_t^h,\ k\left(\theta^{\eta\,h},\, \mu\right) + \displaystyle\int_\omega q_\alpha^{\eta\,h}\,\mu_\alpha = 0 \\[2mm]
\forall\, v_3 \in V_3^h,\ \displaystyle\int_\omega q_\alpha^{\eta\,h}\,\partial_\alpha v_3 = \int_\omega f_3\, v_3 \\[2mm]
\forall\, p \in H_t^h,\ \eta\displaystyle\int_\omega q_\alpha^{\eta\,h}\, p_\alpha - \int_\omega p_\alpha\!\left(\theta_\alpha^{\eta\,h} + \partial_\alpha u_3^{\eta\,h}\right) = 0 \quad.
\end{cases}
$$

This is the standard approximation of (II.42) which is a mixed version of (III.5). For details, the reader is referred to section II.3. Our first result gives the existence and uniqueness of a solution to (III.24) (Theorem III.5). Then general error estimates are obtained in Theorems III.6 and III.7. The main difficulty is then to check conditions (III.22) and (III.23) for an appropriate choice of the spaces W_t^h, V_3^h and H_t^h. This is done in section III.4.2.

Theorem III.5
Let us assume that hypothesis (III.22) is satisfied. Furthermore we suppose that $f_3 \in L^2(\omega)$. Then system III.24 has a unique solution. ∎

Proof of Theorem III.5
Because (III.24) is a finite dimensional linear system, which has the same number of equations and unknowns, it is sufficient to check that the homogeneous system has zero as a unique solution. Setting $\mu = \theta^{\eta\,h}$, $v_3 = u_3^{\eta\,h}$ and $p = q^{\eta h}$ and in (III.24) where we replace f_3 by zero, we deduce (by adding the obtained equations):

$$
k\left(\theta^{\eta h},\, \theta^{\eta h}\right) + \eta\, \|q^{\eta h}\|_{H_t}^2 = 0
$$

and from the coerciveness of the bilinear form, $k(\,.\,,\,.\,)$, on the space W_t :

$$
\theta^{\eta h} = 0
$$

and also:

$$
q^{\eta h} = 0 \quad.
$$

Finally, the third equation (III.24) and the hypothesis (III.22) imply that $u_3^{\eta\,h} = 0$. It is worth noting that the second hypothesis (III.23) has not been used in this Theorem. The condition $f_3 \in L^2(\omega)$ is just used to give a meaning to the term $\displaystyle\int_\omega f_3\, v_3$. Obviously this assumption could be weakened. ∎

Theorem III.6
Let us assume that both hypotheses (III.22) and (III.23) are satisfied. Furthermore we consider a linear continuous form on the space W_t^h which is denoted by y, and a linear continuous form -say x- on H_t^h. We consider an element (z^h, s^h, t^h) in the space $W_t^h \times V_3^h \times H_t^h$ satisfying the system:

(III.25)

$$\begin{cases} \forall \mu \in W_t^h, \; k(z^h, \mu) + \int_\omega t_\alpha^h \mu_\alpha = y(\mu), \\[2mm] \forall v_3 \in V_3^h, \; \int_\omega t_\alpha^h \partial_\alpha v_3 = 0, \\[2mm] \forall p \in H_t^h, \; \eta \int_\omega t_\alpha^h p_\alpha - \int_\omega p_\alpha(z_\alpha^h + \partial_\alpha s^h) = \eta\, x(p). \end{cases}$$

Then there exists a constant – say c – which does not depend on the choice of the spaces W_t^h, V_3^h, H_t^h and such that:

$$\|z^h\|_{W_t} + \|s^h\|_{1,\omega} + \|t^h\|_{M_t^h} + \sqrt{\eta}\,\|t^h\|_{H_t} \le c\,(\|y\| + \sqrt{\eta}\,\|x\|)$$

where:

$$|y(\mu)| \le \|y\| \cdot \|\mu\|_{W_t} \qquad and \qquad |x(p)| \le \|x\| \cdot \|p\|_{H_t} \qquad\blacksquare$$

Proof of Theorem III.6
There are three steps.

Step 1:
Setting $\mu = z^h$, $v_3 = s^h$ and $p = t^h$ in (III.25), and by adding the resulting equations, we obtain:

$$k(z^h, z^h) + \eta\|t^h\|_{H_t}^2 \le \|y\|\,\|z^h\|_{W_t} + \eta\,\|x\|\,\|t^h\|_{H_t}$$

and from Cauchy-Schwarz inequality:

$$\forall\; \alpha > 0 \qquad a b \le \frac{\alpha}{2}a^2 + \frac{1}{2\alpha}b^2,$$

we deduce (taking into account the coerciveness of the bilinear form $k(.,.)$):

(III.26) $c_0\|z^h\|_{W_t}^2 + \eta\|t^h\|_{H_t}^2 \le c_1\left(\|y\|^2 + \eta\|x\|^2\right)$

Step 2:

From hypothesis (III.22) and the third equation (III.25) we deduce that:

(III.27) $\quad c_2 \left\| s^h \right\|_{1,\omega} \leq \displaystyle\sup_{p \in H_t^h} \frac{\displaystyle\int_\omega p_\alpha \, \partial_\alpha s^h}{\left\| p \right\|_{H_t}} \leq \eta \left(\left\| t^h \right\|_{H_t} + \left\| x \right\| \right) + c_3 \left\| z^h \right\|_{W_t}$

and with (III.26) assuming that $\eta < 1$ (and thus: $\eta^2 < \eta$):

(III.28) $\qquad \left\| s^h \right\|_{1,\omega} \leq c_5 \left(\left\| y \right\| + \sqrt{\eta} \left\| x \right\| \right)$

Step 3:

Let us come back to the first two equations (III.25), coupled with hypothesis (III.23). We obtain:

$$c_7 \left\| t^h \right\|_{M_t^h} \leq \sup_{\mu \in W_t^h} \frac{\displaystyle\int_\omega t_\alpha^h \, \mu_\alpha}{\left\| \mu \right\|_{W_t}} + \sup_{v \in V_3^h} \frac{\displaystyle\int_\omega t_\alpha^h \, \partial_\alpha v}{\left\| v \right\|_{1,\omega}}$$

$$\leq \left\| y \right\| + c_7 \left\| z^h \right\|_{W_t}$$

and Theorem III.6 is then deduced from this inequality, (III.27) and (III.28).   ■

From now on, the approximation spaces $W_t^{h\,0}$, $V_3^{h\,0} = V^{h\,0}$, $H_t^{h\,0}$ defined in sections III.4.1.1. and III.4.1.2. are used.

Theorem III.7

Let us assume that both hypotheses (III.22) and (III.23) are satisfied. Let us denote by $\left(\theta^\eta, u_3^\eta, q^\eta \right)$ the solution to (III.5) (for the mixed formulation of the R.M.N. model involving the transverse shear stress q^η). Then we introduce $\pi^3 \theta^\eta$, $\pi^1 u_3^\eta$ and $P^h q^\eta$ which are respectively the interpolates of θ^η, u_3^η, q^η as defined in sections III.4.1.1. and III.4.1.2. . Then there exists a constant c which is independent of the mesh size h and of the small parameter η, and such that:

$$\left\| \theta^\eta - \theta^{\eta\,h} \right\|_{W_t} + \sqrt{\eta} \left\| q^\eta - q^{\eta\,h} \right\|_{H_t} + \left\| u_3^\eta - u_3^{\eta\,h} \right\|_{1,\omega} + \left\| q^\eta - q^{\eta\,h} \right\|_{M_t^h}$$

$$\leq c \left\{ \left\| \theta^\eta - \pi^3 \theta^\eta \right\|_{W_t} + \left\| q^\eta - P^h q^\eta \right\|_{M_t^h} + \left\| u_3^{\eta\,h} - \pi^1 u_3^\eta \right\|_{1,\omega} + \sqrt{\eta} \left\| q^\eta \right\|_{H_t} \right\}$$

where $\theta^{\eta\,h}$, $u_3^{\eta\,h}$ and $q^{\eta\,h}$ are the solutions to the approximate model (III.24). Let us also point out that c is independent of η, (let us point out that π^3 depends on u_3^η).   ■

Proof of Theorem III.7
Let us set first:

$$z^h = \theta^{\eta h} - \pi^3 \theta^\eta, \quad s^h = u_3^{\eta h} - \pi^1 u_3^\eta, \quad t^h = q^{\eta h} - P^h q^\eta,$$

and for arbitrary elements p in the space H_t^{h0} and μ in the space W_t^{h0} we set:

$$x(p) = \int_\omega \left(q_\alpha^\eta - P^h q_\alpha^\eta \right) p_\alpha$$

$$y(\mu) = k\left(\theta^\eta - \pi^3 \theta^\eta, \mu \right) + \int_\omega \left(q_\alpha^\eta - P^h q_\alpha^\eta \right) \mu_\alpha \quad .$$

Then one can observe that $\left(z^h, s^h, t^h \right)$ is solution to the system:

(III.29)
$$\begin{cases} \forall \ \mu \in W_t^{h0}, \ k\left(z^h, \mu \right) + \int_\omega t_\alpha^h \mu_\alpha = y(\mu), \\[2mm] \forall \ v_3 \in V_3^{h0}, \ \int_\omega t_\alpha^h \partial_\alpha v_3 = 0, \\[2mm] \forall \ p \in H_t^{h0}, \ \eta \int_\omega t_\alpha^h p_\alpha - \int_\omega p_\alpha \left(z_\alpha^h + \partial_\alpha s^h \right) = \eta \, x(p) \quad . \end{cases}$$

The basic point for this is to remark that because of Remark III.6 and of the definition of $P^h q_\alpha^\eta$, and $\pi^3 \theta^\eta$ one has:

$$\begin{cases} \forall \ p \in H_t^{h0}, \ \eta \int_\omega p_\alpha \left((\theta_\alpha^\eta - \pi^3 \theta_\alpha^\eta) + \partial_\alpha (u_3^\eta - \pi^1 u_3^\eta) \right) = 0, \\[2mm] \forall \ v_3 \in V_3^{h0} = V^{h0}, \ \int_\omega \left(q_\alpha^\eta - P^h q_\alpha^\eta \right) \partial_\alpha v_3 = 0 \quad . \end{cases}$$

Hence, noticing that, for instance:

$$\left\| \theta^\eta - \theta^{\eta h} \right\|_{W_t} \le \left\| \theta^\eta - \pi^3 \theta^\eta \right\|_{W_t} + \left\| \pi^3 \theta^\eta - \theta^{\eta h} \right\|_{W_t},$$

Theorem III.7 will be a consequence of Theorem III.6 and the following estimates on $\|y\|$ and $\|x\|$.

Estimates on $\|y\|$

First of all, from the definition of $y(\ )$, one has:

$$|y(\mu)| \le c_1 \|\theta^\eta - \pi^3 \theta^\eta\|_{w_t} \|\mu\|_{w_t} + \left| \int_\omega \left(q_\alpha^\eta - P^h q_\alpha^\eta \right) \mu_\alpha \right| .$$

Then let us introduce the approximate of μ by an element of the space $H_t^{h\,0}$. It is denoted by $P^h \mu$ and is such that:

$$P^h \mu = \text{grad } \varphi^h + \text{rot } \psi^h$$

where φ^h and ψ^h are defined by:

$$\begin{cases} \forall \ v \in V^{h\,0}, \ \int_\omega \left(\text{grad } \varphi^h, \ \text{grad } v \right) = \int_\omega (\mu, \ \text{grad } v) \\ \varphi^h \in V^{h\,0}; \end{cases}$$

$$\begin{cases} \forall \ v \in W_0^h, \ \int_\omega \left(\text{rot } \psi^h, \ \text{rot } v \right) = \int_\omega (\mu, \ \text{rot } v) \\ \psi^h \in W_0^h; \end{cases}$$

Furthermore, classical error estimates in finite element methods (see Ciarlet [2] or Raviart-Thomas [3]) lead to the inequalities:

$$\begin{cases} \|\varphi - \varphi^h\|_{1,\omega} \le c\,h\,|\varphi|_{2,\omega} \le c\,h\,\|\mu\|_{w_t} \\ \|\psi - \psi^h\|_{1,\omega} \le c\,h\,|\psi|_{2,\omega} \le c\,h\,\|\mu\|_{w_t} \end{cases}$$

where ψ and φ are defined by:

$$\begin{cases} \mu = \text{grad } \varphi + \text{rot } \psi \\ \varphi \in H^2(\omega) \cap H_0^1(\omega), \ \psi \in H^2(\omega) \cap L_0^2(\omega) \end{cases} .$$

This estimate is much easier or at least more standard than the one we obtained in sections III.4.1.1. and III.4.1.2. This is due to the fact that we have here $\mu_\alpha \in H^1(\omega)$. **But in the approximation of q by $P^h q$ we could only use the regularity** $q_\alpha \in L^2(\omega)$ **and** div $q \in L^2(\omega)$. Let us now notice that from the definitions of $P^h q^\eta$ and $P^h \mu$ one has:

$$\int_\omega \left(q_\alpha^\eta - P^h q_\alpha^\eta \right) \mu_\alpha = \int_\omega \left(q_\alpha^\eta - P^h q_\alpha^\eta \right) \left(\mu_\alpha - P^h \mu_\alpha \right)$$

and thus:

$$\left| \int_\omega \left(q_\alpha^\eta - P^h q_\alpha^\eta \right) \mu_\alpha \right| \leq c\, h \left\| q^\eta - P^h q^\eta \right\|_{H_t} \left\| \mu \right\|_{W_t}$$

where c is a constant which is h- (and μ- !) independent. Finally we proved that:

(III.30) $\|y\| \leq c_2 \left\{ h \left\| q^\eta - P^h q^\eta \right\|_{H_t} + \left\| \theta^\eta - \pi^3 \theta^\eta \right\|_{W_t} \right\}$

c_2 being the maximum of c and c_1.

Estimates on $\|x\|$

It is clear from the definition of the interpolatation of q^η that (only $q_\alpha^\eta \in L^2(\omega)$ is required):

$$\|x\| \leq c \|q^\eta\|_{H_t} \left(= c \sum_{\alpha=1,2} \|q_\alpha^\eta\|_{0,\omega} \right).$$

The Theorem III.7 is then a straightforward consequence of Theorem III.6, equation (III.30) and the triangular inequality.   ∎

III.4.2   *A finite element scheme and error estimates for the R.M.N. model*

This section is a compilation of the previous results (Theorem III.7). First of all let us define the finite element scheme for solving the R.M.N. model:

find $\left(\theta^{\eta h}, u_3^{\eta h}, q^{\eta h} \right) \in W_t^{h0} \times V_3^{h0} \times H_t^{h0}$ such that :

(III.31)
$$\begin{cases} \forall\, \mu \in W_t^{h0},\; k\left(\theta^{\eta h}, \mu \right) + \int_\omega q_\alpha^{\eta h} \mu_\alpha = 0, \\[2mm] \forall\, v_3 \in V_3^{h0},\; \int_\omega q_\alpha^{\eta h} \partial_\alpha v_3 = \int_\omega f\, v_3\,, \\[2mm] \forall\, p \in H_t^{h0},\; \eta \int_\omega q_\alpha^{\eta h} p_\alpha - \int_\omega p_\alpha \left(\theta_\alpha^{\eta h} + \partial_\alpha u_3^{\eta h} \right) = 0. \end{cases}$$

where the finite element spaces $W_t^{h\,0}$ and $H_t^{h\,0}$ were respectively defined in (III.17) and (III.14). Furthermore $V_3^{h\,0}$ is the classical P_1 - Q_1 Lagrange finite element space with continuous functions and homogeneous boundary conditions. Thus, from Theorem III.5, one can ensure that (III.31) has a unique solution. Then from Theorem III.7 and relations (III.12), (III.21), (III.9), (III.10), one deduces that **if the hypotheses** (III.22) **and** (III.23) **are satisfied**, there exists a constant c, which is h- and η-independent and such that:

$$(\text{III.32}) \quad \left\| \theta^\eta - \theta^{\eta h} \right\|_{W_t} + \sqrt{\eta} \left\| q^\eta - q^{\eta h} \right\|_{H_t} + \left\| q^\eta - q^{\eta h} \right\|_{M_t^h} + \left\| u_3^\eta - u_3^{\eta h} \right\|_{1,\omega} \le c \left(h + \sqrt{\eta} \right)$$

From a practical point of view, (III.32) does not indicate how $q^\eta - q^{\eta h}$ behaves when both η and h are small. And this is the practical situation. From the definition of the H^{-1} (ω) norm, one has:

$$\sum_{\alpha=1,2} \left\| q_\alpha^\eta - q_\alpha^{\eta h} \right\|_{-1,\omega} \le c \left\{ \sup_{\varphi=(\varphi_\alpha) \in (H_0^1(\omega))^2} \frac{\left| \int_\omega \left(q_\alpha^\eta - q_\alpha^{\eta h} \right) \varphi_\alpha \right|}{\sum_{\alpha=1,2} \left\| \varphi_\alpha \right\|_{1,\omega}} \right\}.$$

But, let us remark that from the definition of q^η and $q^{\eta h}$:

$$\forall \, \mu \in W_t^{h\,0}, \quad \int_\omega \left(q_\alpha^{\eta h} - q_\alpha^\eta \right) \mu_\alpha = - \, k \left(\theta^{\eta h} - \theta^\eta, \mu \right).$$

Therefore for any φ^h lying in the space $W_t^{h\,0}$, which is an internal approximation of $\left(H_0^1 \right)^2$, one has:

$$\left\| q_\alpha^\eta - q_\alpha^{\eta h} \right\|_{-1,\omega} \le \sup_{\varphi=(\varphi_\alpha) \in (H_0^1(\omega))^2} \left[\frac{\int_\omega \left(q_\alpha^\eta - q_\alpha^{\eta h} \right) \left(\varphi_\alpha - \varphi_\alpha^h \right)}{\sum_{\alpha=1,2} \left\| \varphi_\alpha \right\|_{1,\omega}} + \frac{k(\theta^{\eta h} - \theta^\eta, \varphi^h)}{\sum_{\alpha=1,2} \left\| \varphi_\alpha^h \right\|_{1,\omega}} \right]$$

and from (III.32) and Theorem III.2, choosing for φ^h the $L^2(\omega)$-projection of φ onto the space $W_t^{h\,0}$, we deduce that there exists a constant c_0 which is both h- and η-independent, and such that for η small enough, one has:

$$\left\| q_\alpha^\eta - q_\alpha^{\eta h} \right\|_{-1,\omega} \le c_0 \left(h + \sqrt{\eta} \right).$$

Obviously this estimate is a little bit difficult to observe numerically (i.e. from a computational point of view). This is due to the fact that negative Sobolev spaces are not easy to manipulate. For instance the norm cannot be estimated locally (see R. Adams [23]). But unfortunately it is quite impossible to improve this result (III.33). This is because $H^{-1}(\omega)$ is the right space for the transverse shear stress q_α^η when η tends to zero.

III.4.2.1   Checking the hypotheses (III.22) and (III.23) with the spaces $W_t^{h\,0}, V_3^{h\,0}$ and $H_t^{h\,0}$

i) Let us first consider the inequality (III.22). The gradient of a function v_3 lying in the space $V_3^{h\,0}$ is constant on a triangle, and equal to: $(a + b\,x_2, c + b\,x_1)$ on a quadrilateral (on the reference element $\hat{K}$). Hence it is possible to choose: $p_\alpha = \partial_\alpha u_3$ in order to check hypothesis III.22. Hence III.22 is quite obvious to prove.

ii) Let us now check of hypothesis III.23. From the same considerations as in (III.18), for any element $p = (p_\alpha)$ in the space $H_t^{h\,0}$, one can choose an element μ^0 in the space $W_t^{h\,0}$ such that :

$$\begin{cases} \displaystyle\int_\omega p_\alpha \mu_\alpha^0 = \int_\omega p_\alpha p_\alpha \;, \\[2mm] \left\| \mu^0 \right\|_{H_t} \leq c_1 \left\| p \right\|_{H_t} \;. \end{cases}$$

This is just obtained by adjusting the internal degrees of freedom of μ^0. With another respect, the inverse inequality in finite element methods gives (see for instance Ciarlet [2]):

$$\forall\, \mu \in W_t^{h\,0}, \; \left\| \mu \right\|_{W_t} \leq \frac{c_2}{h} \left\| \mu \right\|_{H_t} \;.$$

Thus we obtain:

$$\forall\, p \in H_t^{h\,0}, \quad \sup_{\mu \in W_t^{h\,0}} \frac{\displaystyle\int_\omega p_\alpha \mu_\alpha}{\left\| \mu \right\|_{W_t}} \geq \frac{h}{c_2} \sup_{\mu \in W_t^{h\,0}} \frac{\displaystyle\int_\omega p_\alpha \mu_\alpha}{\left\| \mu \right\|_{H_t}}$$

and choosing $\mu = \mu^0$ which has been defined previously :

$$\forall\, p \in H_t^{h\,0}, \quad \sup_{\mu \in W_t^{h\,0}} \frac{\displaystyle\int_\omega p_\alpha \mu_\alpha}{\left\| \mu \right\|_{W_t}} \geq c_3\, h\, \left\| p \right\|_{H_t}$$

which establishes the validity of (III.23) by setting $v_3 = 0$.

III.4.3   *Practical aspects in solving the R.M.N. finite element model*

The last equation (III.31) enables one to eliminate locally (i.e. element by element) the transverse shear stress $q_\alpha^{\eta h}$. In the case of a triangle, $q_\alpha^{\eta h}$ is piecewise constant. Thus $\eta \, q_\alpha^{\eta h}$ appears to be the L^2 (K) projection of $\theta_\alpha^{\eta h} + \partial_\alpha u_3^{\eta h}$ on the constants. Then the model (III.31) can be interpreted as follows. Let us consider the energy associated with (III.5):

$$ J\left(\theta , u_3\right) = \tfrac{1}{2} k\left(\theta , \theta\right) + \frac{1}{2\eta} \int_\omega \sum_{\alpha = 1 , 2} \left(\theta_\alpha + \partial_\alpha u_3\right)^2 - \int_\omega f_3 \, u_3 . $$

When the mesh T^h contains only triangles, the approximate model is:

$$ \min_{(\mu , v_3) \, \in \, W_t^{h0} \times V_3^{h0}} J^h\left(\mu , v_3\right) $$

where approximate functional J^h is defined by :

(III.34) $$ J^h\left(\mu , v_3\right) = \tfrac{1}{2} k\left(\mu , \mu\right) + \frac{1}{2\eta} \sum_{K \in T^h} \left[\int_K \left(\mu_\alpha + \partial_\alpha v_3\right) \right]^2 - \int_\omega f_3 \, v_3 . $$

Thus the definition of J^h can be seen as a modification of J obtained by a reduced integration on the transverse shear energy (i.e. the penalty term). This phenomenon has been widely discussed by various authors for quadrilaterals. But it is different from what happens here, because (III.34) corresponds to triangles. As a matter of fact, it is not possible in the present approach to cancel the internal degrees of freedom for the rotation θ, as they play a basic role. Nevertheless, the scheme presented here looks like the one suggested by T. Belytschko and M. Stolarski [24], even if it leads to a different model. Let us now consider the general case of a mesh with both triangles and rectangles. It could be also possible to use deformed quadrilaterals with the standard changes. Then it is possible to define a closed subspace U^{h0} of $W_t^{h0} \times V_3^{h0}$ such that: $\left(\theta , u_3\right) \in U^{h0}$ if :

(III.35) $$ \forall \, p \in H_t^{h0} , \quad \int_\omega \left(\theta_\alpha + \partial_\alpha u_3\right) p_\alpha = 0 $$

and we recall that H_t^{h0} is defined in (III.14). Then an approximation of (III.31) consists in finding an element $\left(\theta^{h\,0},\, u^{h}_3{}^{0}\right) \in \boldsymbol{U}^{h\,0}$ such that:

(III.36) $\forall\, (\mu,\, v_3)\, \cdot \in \boldsymbol{U}^{h\,0},\ k\left(\theta^{h\,0},\mu\right) = \displaystyle\int_{\omega} f_3\, v_3$

which can be interpreted as a discretized Kirchhoff-Love model where the exact relation

$$\theta_\alpha + \partial_\alpha u_3 = 0 \quad \text{on} \quad \omega \ \text{for}\ \alpha = 1,2$$

has been weakened. Obviously error estimates are no more valid for this new model. But it seems clear that the accuracy is certainly not as good as the one obtained with (III.31) because $\boldsymbol{U}^{h\,0}$ is smaller than $W^{h\,0}_t \times V^{h\,0}_3$. In addition, the last equation (III.31) suggests to set $q^\eta = 0$. So this discrete Kirchhoff-Love assumption, which is "a great temptation" because of the simplicity of the model obtained, is certainly not well adapted to situations where the transverse shear stress plays an important role. Concerning the practical aspects of the implementation of (III.36), let us mention how simple it is. The degrees of freedom are those of θ and u_3 at the vertices of the mesh. The internal degrees of freedom of θ are fixed from relation (III.35). Hence a two-dimensional elasticity solver is sufficient for solving (III.36).

III.4.4   About the famous QUAD 4 element

Most finite element codes for structural analysis have the QUAD 4 element. It is based also on the R.M.N. formulation, but has no mathematical justification. We are going to prove that this model is almost equivalent to the one we have studied in this paragraph III.4. For sake of clarity, let us recall briefly the QUAD 4 formulation.

III.4.4.1   The QUAD 4 of R. MacNeal - T. Hughes

Let us introduce the following approximation spaces for respectively θ^η, $u^{\,\eta}_3$ and q^η. First of all θ^η should be in the space:

(III.37) $W^{h\,1}_t = \left\{\mu\,|\,\mu = (\mu_\alpha) \in W_t;\ \forall\ K \in Q^h, \mu_{\alpha|K} \in Q_1\right\}.$

One important feature is that T^h contains only rectangles (or quadrilaterals with the classical changes concerning the definition of the shape functions, which are Q_1 on the reference element and not on the element K). Then the space $V^{h\,0}_3$ is unchanged: $V^{h\,1}_3 = V^{h\,0}_3$. Finally, $H^{h\,1}_t$ is also unchanged with respect to $H^{h\,0}_t$ and is defined by (as far as Q^h only contains quadrilaterals):

(III.38) $H_t^{h\,1} = \left\{ p \,|\, p = (p_\alpha) \in \left(L^2(\omega)\right)^2 ; \ \forall \ K \in Q^h , p_{\alpha|K} \in Q_0 \right\}$.

Then the QUAD 4 model consists in finding an element ($\theta^{\eta\,h\,1} , u_3^{\eta\,h\,1} , q^{\eta\,h\,1}$) lying in the space $W_t^{h\,1} \times V_3^{h\,1} \times H_t^{h\,1}$, such that:

(III.39) $\begin{cases} \forall \ \mu \in W_t^{h\,1} , \ k\left(\theta^{\eta\,h\,1} , \mu\right) + \displaystyle\int_\omega q_\alpha^{\eta\,h\,1} \mu_\alpha = 0 , \\[2mm] \forall \ v_3 \ \in V_3^{h\,1} , \ \displaystyle\int_\omega q_\alpha^{\eta\,h\,1} \partial_\alpha v_3 = \int_\omega f_3\, v_3 , \\[2mm] \forall \ p \in H_t^{h\,1} , \ \eta \displaystyle\int_\omega q_\alpha^{\eta\,h\,1} p_\alpha - \int_\omega p_\alpha\left(\theta_\alpha^{\eta\,h\,1} + \partial_\alpha u_3^{\eta\,h\,1}\right) = 0 . \end{cases}$

We do not discuss directly the validity of this model (which is not exactly justified as we see later on). But the existence and uniqueness of a solution is quite obvious from Theorem III.5. We leave it to the reader. Let us rather examine how this model is close to the one we gave in (III.3.1.).

Remark III.7

The main difference between (III.31) and (III.39) is the elimination of the internal degrees of freedom for elements of both $W_t^{h\,1}$ and $H_t^{h\,1}$.   ■

III.4.4.2   A different formulation for the R.M.N. finite element model when T^h *is made of quadrilaterals*

The first step consists in splitting the approximation space $W_t^{h\,0}$ into two subspaces. One, denoted $W_t^{h\,i}$, corresponds to the internal degrees of freedom (4 for each rectangle). The other one is isomorphic to $W_t^{h\,1}$, but with different shape functions; it is denoted by $W_t^{h\,s}$ (s means Serendipity). Let us make explicit how $W_t^{h\,s}$ is built. First of all, let N^i (i = 1 , 2) be one of the two shape functions associated with a summit of K. Then we set:

$$N^{i\,s} = N^i - \sum_{\lambda = 1,2} \xi_\lambda^i L_K^\lambda \quad \text{for } i = 1,2 ,$$

where ξ_λ^i is a real number to be determined, and L_K^i one of the two functions (i = 1 , 2) constructed with the bubble B_K (cf. III.17). The coefficients ξ_λ^i are then computed by the equations (for i = 1, 2 and k being the bilinear form used up to now):

$$\begin{cases} \xi_1^i \, k\left(L_K^1, L_K^1\right) + \xi_2^i \, k\left(L_K^2, L_K^1\right) = k\left(N^i, L_K^1\right), \\ \xi_1^i \, k\left(L_K^1, L_K^2\right) + \xi_2^i \, k\left(L_K^2, L_K^2\right) = k\left(N^i, L_K^2\right). \end{cases}$$

This system can easily be solved because k is obviously elliptic on $\left[H_0^1(K)\right]^2$; and $L_K^\lambda \in H_0^1(K)$. The space $W_t^{h\,s}$ is then built with shape functions which are N^{is} instead of N^i. A classical exercise left to the reader shows that $N^i = N^{i\,s}$ if K is a rectangle and therefore $W_t^{h\,1} = W_t^{h\,s}$ if T^h only contains rectangles. This remark is very important for the rest of the section. There are two other degrees of freedom associated with the functions of the space R_{KQ} (see (III.17)), say M_K^i for $i = 1 , 2$. But, if we set:

$$m_{\alpha\beta}^i = \frac{2\,E\,\varepsilon^3}{3\left(1 - v^2\right)} \left\{(1 - v)\,\gamma_{\alpha\beta}\left(L_K^i\right) + v\,\gamma_{\lambda\lambda}\left(L_K^i\right)\delta_{\alpha\beta}\right\}$$

then, the bubble being globally C^2 on each K :

$$k\left(L_K^i, M_K^j\right) = -\int_K \partial_\beta \, m_{\beta\alpha}^i \left(M_K^j\right)_\alpha \ .$$

An easy calculation shows that this term is zero, using the skew-symmetrical properties, and thus:

$$k\left(L_K^i, M_K^j\right) = 0$$

It is also quite easy to check directly that:

$$k\left(N^i, M_K^j\right) = -\int_K \partial_\beta \, m_{\beta\alpha}^i \left(M_K^j\right)_\alpha = 0$$

where:

$$m_{\alpha\beta}^i = \frac{2\,E\,\varepsilon^3}{3\left(1 - v^2\right)} \left\{(1 - v)\,\gamma_{\alpha\beta}\left(N^i\right) + v\,\gamma_{\lambda\lambda}\left(N^i\right)\delta_{\alpha\beta}\right\} \ .$$

Hence $W_t^{h\,s}$ and $W_t^{h\,i}$ are conjugate with respect to the bilinear form $k\left(.,.\right)$. Let us set:

(III.40) $\theta^{\eta h} \in W_t^{h\,0} :\ \theta^{\eta h} = \theta^{h\,1} + \theta^{h\,2}$ where $\theta^{h\,1} \in W_t^{h\,s}, \theta^{h\,2} \in W_t^{h\,i}$

and thus the first equation (III.31) can also be written:

(III.41) $\forall \mu \in W_t^{hs}, \quad k\left(\theta^{h1}, \mu\right) + \int_\omega q_\alpha^{\eta h} \mu_\alpha = 0,$

(III.42) $\forall \mu \in W_t^{hi}, \quad k\left(\theta^{h2}, \mu\right) + \int_\omega q_\alpha^{\eta h} \mu_\alpha = 0.$

The equation (III.42) permits one to eliminate locally (ie. element by element) θ^{h2} as a function of $q^{\eta h}$. For further details, we refer the reader to the calculations performed in paragraph III.4.1.2. The third relation (III.31) can then be written:

(III.43) $\forall \ p \in H_t^{h0}, \quad \eta \int_\omega \left[q_\alpha^{\eta h} - \frac{1}{\eta} \theta^{h2} (q^{\eta h}) \right] p_\alpha - \int_\omega p_\alpha \left(\theta_\alpha^{h1} + \partial_\alpha u_3^{\eta h} \right) = 0.$

and the R.M.N. finite element method can be formulated as follows:

(III.44)
$$
\begin{cases}
\text{find} \left(\theta^{h1}, u_3^{\eta h}, q^{\eta h} \right) \in W_t^{hs} \times V_3^{h0} \times H_t^{h0} \quad \text{such that :} \\[2mm]
\forall \ \mu \in W_t^{hs}, \quad k\left(\theta^{h1}, \mu\right) + \int_\omega q_\alpha^{\eta h} \mu_\alpha = 0, \\[2mm]
\forall \ v_3 \in V_3^{h0}, \quad \int_\omega q_\alpha^{\eta h} \partial_\alpha v_3 = \int_\omega f_3 \, v_3, \\[2mm]
\forall \ p \in H_t^{h0}, \quad \eta \int_\omega T_\alpha (q^{\eta h}) p_\alpha - \int_\omega p_\alpha \left(\theta_\alpha^{h1} + \partial_\alpha u_3^{\eta h} \right) = 0.
\end{cases}
$$

where:

(III.45) $\qquad\qquad T_\alpha (q^{\eta h}) = q_\alpha^{\eta h} - \frac{1}{\eta} \theta_\alpha^{h2} (q^{\eta h})$

$\theta_\alpha^{h2} (q^{\eta h})$ being calculated from (III.42). In order to simplify the numerical implementation of the model (III.44), one can suggest to approximate the terms:

$$\int_\omega p_\alpha \mu_\alpha , \int_\omega p_\alpha \partial_\alpha v_3 \text{ and } \int_\omega T_\alpha (q^{\eta h}) p_\alpha$$

by a simplified integration. Then we set for these terms:

(III.46) $\qquad\qquad \int_\omega f \cong \text{meas} \, (K) \ f (G_K)$

where f is a function, G_K the center of gravity of K and mes (K), the surface of K. In such a case, the skew-symmetrical terms of $q^{\eta h}$ are eliminated because they are zero at the center of gravity of each element K of the mesh T^h. For the contributions of $\theta_\alpha^{h\,2}(q^{\eta h})$, only those coming from the bubble function are kept (ie. we forget those coming from the functions R_{KQ}; see III.17). Finally, if the quadrilaterals are rectangles, $W_t^{h\,s} = W_t^{h\,1}$ and the model (III.44), where we have used the reduced integration previously mentioned, is "almost" identical to the QUAD 4 formulation. The only difference is the definition of the transverse shear energy, which is:

(III.47)
$$\frac{\eta}{2} \int_\omega q_\alpha^{\eta h}\, q_\alpha^{\eta h}$$

for the QUAD 4, and:

(III.48)
$$\frac{\eta}{2} \int_\omega T_\alpha(q^{\eta h})\, q_\alpha^{\eta h}$$

for the R.M.N. finite element model.

As a matter of fact, this difference can certainly be neglected for η small enough (except may be for the transverse shear stress itself). Obviously, according to mechanical considerations, it is much better to use the right three-dimensional expression of this shear energy. This is the way it is done in structural analysis codes, and therefore this slight difference between the QUAD 4 formulation and the one obtained here can be interpreted by a modification of the three-dimensional shear energy which is quite meaningless on the components θ^η and u_3^η.

But one has to be more careful concerning the interpretation of the transverse shear stress q^η, for which III.48 is certainly better. Let us point out here that this discussion has been at the origin of a very large number of papers from both mathematicians and engineers. But to our best knowledge, a convincing justification concerning which one of the two expressions is the more suitable for numerical purpose, has never been formulated in a convincing manner. This question is nevertheless a fundamental one, as far as one is concerned with a mechanical problem in which the transverse shear stress is meaningful. This is the case for delamination of thin laminated composite plates or shells. This problem is extensively studied in the fifth chapter of this book.

III.5   Numerical approximation of the mixed formulation for a bending plate

The goal of this section is to analyze several finite element schemes based on the "Natural Duality" formulation introduced in Chapter II, section II.4. Let us first recall briefly the variational formulation we obtained. We assume hereafter (again) that the boundary of ω is simply connected. The generalization to multi-connected domain is given in section II.4.8

First of all we introduced a potential function – say φ – such that:

(III.49)

$$
\begin{cases}
- \Delta \varphi = f_3 \text{ on } \omega , \\[2mm]
\varphi = 0 \text{ on } \gamma_0 \cup \gamma_1 , \\[2mm]
\dfrac{\partial \varphi}{\partial b} = 0 \text{ on } \gamma_2 .
\end{cases}
$$

The variational formulation of this classical model is the following:

$$
\begin{cases}
\text{find } \varphi \in V = \left\{ v \mid v \in H^1(\omega), \ v = 0 \text{ on } \gamma_0 \cup \gamma_1 \right\}, \\[2mm]
\forall \ v \in V, \ \displaystyle\int_\omega \text{grad } \varphi \cdot \text{grad } v = \int_\omega f_3 \, v .
\end{cases}
$$

Let us define a finite dimensional subspace V^h of V, generated by a first-order finite element method (i.e. C^0 and piecewise linear or bilinear functions). The approximate problem is defined by:

(III.50)

$$
\begin{cases}
\text{find } \varphi^h \in V^h , \text{ such that:} \\[2mm]
\forall \ v \in V^h, \ \displaystyle\int_\omega \text{grad } \varphi^h \cdot \text{grad } v = \int_\omega f_3 \, v ,
\end{cases}
$$

and from classical results (see Ciarlet [2] or Raviart-Thomas [3]), the error between φ and φ^h is given by:

(III.51)

$$
\begin{cases}
\left\| \varphi - \varphi^h \right\|_{1, \omega} \leq c \, h \, |\varphi|_{2, \omega} \\[2mm]
\left\| \varphi - \varphi^h \right\|_{0, \omega} \leq c \, h^2 \, |\varphi|_{2, \omega}
\end{cases}
$$

(see section I.1. for a brief recall on the notations). The above estimates are meaningful only if

$\varphi \in H^2(\omega)$, which requires that $f_3 \in L^2(\omega)$, but this is not sufficient).

The rotation θ which represents the partial derivatives of the deflection u_3 (regardless of the sign) is solution to the mixed model explicited hereafter:

$$\text{find } (\theta, \psi, c) \in W_t \times L_0^2(\omega) \times \mathbb{R}^N \text{ such that:}$$

(III.52)
$$\begin{cases} \forall \mu \in W_t, \ k(\theta, \mu) + b(\Lambda, \mu) = g(\mu) \\ \forall \Xi \in M, \ b(\Xi, \theta) = 0; \end{cases}$$

with the notations $M = L_0^2(\omega) \times \mathbb{R}^N$ and $\Lambda = (\psi, c)$. Other notations used are those introduced in section II.3.2. Let us recall that:

$$g(\mu) = - \int_\omega \text{grad } \varphi \cdot \mu$$

and

$$\forall \Lambda \in M, \ \sup_{\mu \in W_t} \frac{b(\Lambda, \mu)}{\mu_{W_t}} \geq \mu_M.$$

First of all, let us point out that the approximate model for θ and Λ is defined with :

$$g^h(\mu) = - \int_\omega \text{grad } \varphi^h \cdot \mu,$$

instead of $g(\mu)$. Our first result concerns the error between g and g^h. Obviously one has:

(III.53)
$$g^h(\mu) - g(\mu) \leq c h \ \varphi_{2,\omega} \ \mu_{W_t}$$

but it is important to notice that an improvement can be obtained in the next Theorem III.8 (i.e. h^2 instead of h). Then the approximate model to (III.52) is:

(III.54)
$$\text{find } (\theta^h, \Lambda^h) \in W_t^h \times M^h \text{ such that:}$$
$$\begin{cases} \forall \mu \in W_t^h, \ k(\theta^h, \mu) + b(\Lambda^h, \mu) = g^h(\mu) \\ \forall \Xi \in M^h, \ b(\Xi, \theta^h) = 0; \end{cases}$$

where W_t^h and M^h are finite dimensional subspaces of W_t and M, which are explicited in the following (there are several possibilities).

Theorem III.8

Let us assume that for any element q in the space $H^{1/2}(\gamma_2)$, *the element z of the space*

$$V = \left\{v \mid v \in H^1(\omega), \ v = 0 \text{ on } \gamma_0 \cup \gamma_1\right\},$$

such that:

$$\forall \ v \in V, \ \int_\omega \operatorname{grad} z \cdot \operatorname{grad} v = \int_{\gamma_2} q \, v$$

satisfies the regularity assumption $z \in H^2(\omega)$ *and furthermore:*

$$\|z\|_{2,\omega} \le c_0 \, h \, \|q\|_{1/2,\gamma_2} .$$

Then there exists a constant c_4 *wich depends on* $\|\varphi\|_{2,\omega}$ *and such that:*

$$\left| g(\mu) - g^h(\mu) \right| \le c_4 \, h^2 \|\mu\|_{W_t}$$

(obviously c_4 *is assumed to be independent of* h *and of* μ). $\blacksquare$

Proof of Theorem III.8

Let us first notice that for any function μ in the space W_t one has:

$$g(\mu) - g^h(\mu) = \int_\omega \left(\operatorname{grad} \varphi^h - \operatorname{grad} \varphi\right) \cdot \mu = \int_{\gamma_2} \left(\varphi^h - \varphi\right) \mu_\alpha \, b_\alpha - \int_\omega \left(\varphi^h - \varphi\right) \operatorname{div} \mu .$$

Hence, using the continuity of the trace mapping, we obtain:

$$\left| g(\mu) - g^h(\mu) \right| \le c \left\{ \left\| \varphi^h - \varphi \right\|_{-1/2,\gamma_2} + \left\| \varphi^h - \varphi \right\|_{0,\omega} \right\} \|\mu\|_{W_t} .$$

We know from classical results (see (III.51)) that:

$$\left\| \varphi - \varphi^h \right\|_{0,\omega} \le c \, h^2 \|\varphi\|_{2,\omega} .$$

It remains to prove the same thing for $\varphi - \varphi^h$ in the space $H_{00}^{-1/2}(\gamma_2)$, dual space of $H_{00}^{1/2}(\gamma_2)$, (which is the only difficulty). If we assume the regularity which enables one to use the integrals instead of the duality product, we notice that :

$$\left\| \varphi - \varphi^h \right\|_{-1/2, \gamma_2} = \sup_{q \in H_{00}^{1/2}(\gamma_2)} \frac{\displaystyle\int_{\gamma_2} q \left(\varphi - \varphi^h \right)}{\left\| q \right\|_{1/2, \gamma_2}} .$$

But from the definition of the function z, one has:

$$\int_{\gamma_2} q \left(\varphi - \varphi^h \right) = \int_{\omega} \text{grad } z \cdot \text{grad} \left(\varphi - \varphi^h \right)$$

or else, if z^h is the V^h interpolate of z :

$$\int_{\gamma_2} q \left(\varphi - \varphi^h \right) = \int_{\omega} \text{grad} \left(z - z^h \right) \cdot \text{grad} \left(\varphi - \varphi^h \right)$$

hence:

$$\left| \int_{\gamma_2} q \left(\varphi - \varphi^h \right) \right| \leq \left\| z - z^h \right\|_{1, \omega} \left\| \varphi - \varphi^h \right\|_{1, \omega}$$

$$\leq c_1 h^2 \left| z \right|_{2, \omega} \left| \varphi \right|_{2, \omega}$$

$$\leq c_2 h^2 \left\| q \right\|_{1/2, \gamma_2} \left| \varphi \right|_{2, \omega} .$$

Finally:

$$\left\| \varphi - \varphi^h \right\|_{-1/2, \gamma_2} \leq c_2 h^2 \left| \varphi \right|_{2, \omega}$$

which completes the proof of Theorem III.8.   ∎

III.5.1   *General error estimate between* (θ, Λ) *and* (θ^h, Λ^h)

We use the following hypothesis.

H$_1$ - *There exists a strictly positive constant which is independent of* h – *say* c_0 – *such that:*

$$\boxed{\quad \forall \, \Lambda \in M^h, \quad \sup_{\mu \in W_t^h} \frac{b(\Lambda, \mu)}{\left\| \mu \right\|_{W_t}} \geq c_0 \left\| \Lambda \right\|_M \quad}$$

This enable us to prove the following result:

Theorem III.9

Assuming that the hypothesis H_1 is satisfied, there exists a constant c such that:

$$\|\theta - \theta^h\|_{W_t} + \|\Lambda - \Lambda^h\|_M \le c\left[\inf_{(\mu, \Xi) \in W_t^h \times M^h} \left\{ \|\theta - \mu\|_{W_t} + \|\Lambda - \Xi\|_M \right\} + h^2 \right]$$

where (θ, Λ), (respectively (θ^h, Λ^h)), denotes the solution to (III.52), (respectively (III.54)).   ∎

Proof of Theorem III.9 (which is quite classical for mixed methods)

Let us first notice that there exists a strictly positive constant c_0 such that for any element μ in the space W_t^h one has:

$$c_0\|\theta - \theta^h\|_{W_t}^2 \le k(\theta - \theta^h, \theta - \theta^h) = k(\theta - \theta^h, \theta - \mu) + k(\theta - \theta^h, \mu - \theta^h)$$

or else using Cauchy-Schwarz inequality and (III.52) - (III.54):

$$c_0'\|\theta - \theta^h\|_{W_t}^2 \le c_1 \left\{ \|\theta - \mu\|_{W_t}^2 + |g(\mu - \theta^h) - g^h(\mu - \theta^h)| \right\} - b(\Lambda - \Lambda^h, \mu - \theta^h) .$$

Theorem III.8 and the triangular inequality lead to:

$$|g(\mu - \theta^h) - g^h(\mu - \theta^h)| \le c h^2 \left[\|\mu - \theta\|_{W_t} + \|\theta - \theta^h\|_{W_t} \right]$$

and thus, one has for h small enough:

(III.55) $$\|\theta - \theta^h\|_{W_t}^2 \le c_2 \left\{ \|\theta - \mu\|_{W_t}^2 + h^4 \right\} + |b(\Lambda - \Lambda^h, \mu - \theta^h)| .$$

But for any element Ξ in the space M^h, one has:

$$|b(\Lambda - \Lambda^h, \mu - \theta^h)| \le |b(\Lambda - \Xi, \mu - \theta^h)| + |b(\Xi - \Lambda^h, \mu - \theta^h)|$$

and because of equations (III.54) and (III.52):

(III.56)
$$\begin{aligned} |b(\Lambda - \Lambda^h, \mu - \theta^h)| &\le |b(\Lambda - \Xi, \mu - \theta^h)| + |b(\Xi - \Lambda^h, \mu - \theta)| \\ &\le c_3 \|\Lambda - \Xi\|_M \left\{ \|\mu - \theta\|_{W_t} + \|\theta - \theta^h\|_{W_t} \right\} \\ &+ c_4 \|\Lambda - \Lambda^h\|_M \|\mu - \theta\|_{W_t} . \end{aligned}$$

Finally, from (III.55) and (III.56) we deduce (using: $ab \leq \frac{\xi}{2} a^2 + \frac{1}{2\xi} b^2$):

(III.57) $\|\theta - \theta^h\|_{W_t}^2 \leq c_5 \left\{ \|\theta - \mu\|_{W_t}^2 + h^4 + \|\Lambda - \Xi\|_M^2 + \|\Lambda - \Lambda^h\|_M \|\mu - \theta\|_{W_t} \right\}$.

From hypothesis $\mathbf{H_1}$, we deduce that:

$$\|\Lambda - \Lambda^h\|_M \leq \|\Lambda - \Xi\|_M + c_6 \sup_{\mu \in W_t^h} \frac{b\left(\Xi - \Lambda^h, \mu\right)}{\|\mu\|_{W_t}}$$

or else, with the continuity of the bilinear form b (. , .) and the triangular inequality :

(III.58) $\|\Lambda - \Lambda^h\|_M \leq c_7 \|\Lambda - \Xi\|_M + c_6 \sup_{\mu \in W_t^h} \frac{b\left(\Lambda - \Lambda^h, \mu\right)}{\|\mu\|_{W_t}}$.

Then, using equations (III.52)-(III.54) and Theorem III.8, we deduce:

$$\sup_{\mu \in W_t^h} \frac{b\left(\Lambda - \Lambda^h, \mu\right)}{\|\mu\|_{W_t}} \leq c_8 h^2 + \sup_{\mu \in W_t^h} \frac{k\left(\theta - \theta^h, \mu\right)}{\|\mu\|_{W_t}}$$

or else:

(III.59) $$\sup_{\mu \in W_t^h} \frac{b\left(\Lambda - \Lambda^h, \mu\right)}{\|\mu\|_{W_t}} \leq c_8 h^2 + \|\theta - \theta^h\|_{W_t}$$.

Finally, (III.58)-(III.59) give for any $\Xi \in \mathbf{M}^h$:

(III.60) $\|\Lambda - \Lambda^h\|_M \leq c_9 \left\{ \|\theta - \theta^h\|_{W_t} + \|\Lambda - \Xi\|_M + h^2 \right\}$.

Combining (III.60) and (III.57) leads finally to the estimate of Theorem III.9.   ∎

The error estimates between θ and θ^h on the one hand, Λ and Λ^h on the other hand, will be directly deduced from Theorem III.4 for various finite element schemes in the following sections. But obviously, it will be necessary to check hypothesis $\mathbf{H_1}$. In order to prove a better accuracy on $\theta - \theta^h$ in the $(L^2(\omega))^2$ norm, we shall use the following technical result which is an extension to mixed a formulation of Nitsche method introduced for elliptic problems (see Ciarlet [2] or Raviart-Thomas [3]).

Let us first consider for any element $\varphi = (\varphi_\alpha)$ in the space $(L^2(\omega))^2$ the following mixed problem:

$$(III.61) \quad \begin{cases} \text{find } (z, \chi) \in W_t \times M \text{ such that:} \\[2mm] \forall\ \mu\ \in\ W_t,\ k(z, \mu) + b(\chi, \mu) = \int_\omega \varphi_\alpha \mu_\alpha \\[2mm] \forall\ \Xi \in M,\ b(\Xi, z) = 0 \ . \end{cases}$$

It is clear that (III.61) has a unique solution.

Then let (z^h, χ^h) be the unique solution to the approximate model defined by:

$$(III.62) \quad \begin{cases} \text{find } (z^h, \chi^h) \in W_t^h \times M^h \text{ such that:} \\[2mm] \forall\ \mu\ \in\ W_t^h,\ k(z^h, \mu) + b(\chi^h, \mu) = \int_\omega \varphi_\alpha \mu_\alpha\ , \\[2mm] \forall\ \Xi \in M^h,\ b(\Xi, z^h) = 0 \ . \end{cases}$$

First of all, setting $\Xi = \chi^h$ and $\mu = z^h$ we deduce the "a priori" estimate:

$$c\|z^h\|_{W_t}^2 \leq k(z^h, z^h) = \int_\omega \varphi_\alpha z_\alpha^h \leq \left[\sum_{\alpha = 1, 2} \|\varphi_\alpha\|_{0, \omega} \|z^h\|_{W_t} \right]$$

or else:

$$(III.63) \qquad \|z^h\|_{W_t} \leq (1/c) \left[\sum_{\alpha = 1, 2} \|\varphi_\alpha\|_{0, \omega}^2 \right]^{1/2} .$$

The same proof as the one used in Theorem III.9 enables one to establish that there exists a constant c such that (assuming that $\mathbf{H_1}$ is true and c being a new constant):

$$(III.64) \quad \|z - z^h\|_{W_t} + \|\chi - \chi^h\|_M \leq c \inf_{(\mu, \Xi) \in W_t^h \times M^h} \left\{ \|z - \mu\|_{W_t} + \|\chi - \Xi\|_M \right\} .$$

It is interesting to point out that the term h^2 has disappeared because φ_α is the same in this case for both the continuous and the approximate systems.

Theorem III.10

Let us assume that hypothesis $\mathbf{H_1}$ is true. Then there exists a constant c such that:

$$\sum_{\alpha=1,2} \left\| \theta_\alpha - \theta_\alpha^h \right\|_{0,\omega} \le c \left[\sup_{\varphi \in (L^2(\omega))^2} \left\{ \inf_{(\mu,\Xi) \in W_t^h \times M^h} \frac{\| z - \mu \|_{W_t} + \| \chi - \Xi \|_M}{\sum_{\alpha=1,2} \| \varphi_\alpha \|_{0,\omega}} \right\} \left\| \theta - \theta^h \right\|_{W_t} + h^2 \right]$$

where (z, χ) *is solution to the mixed model* (III.61). *In this estimate* θ, *respectively* θ^h, *is solution to* (III.52), *respectively* (III.54). ∎

Proof of Theorem III.10

First of all, one has:

$$\sum_{\alpha=1,2} \left\| \theta_\alpha - \theta_\alpha^h \right\|_{0,\omega} = \sup_{\varphi \in (L^2(\omega))^2} \frac{\displaystyle\int_\omega \varphi_\alpha \left(\theta_\alpha - \theta_\alpha^h \right)}{\sum_{\alpha=1,2} \| \varphi_\alpha \|_{0,\omega}} .$$

Then from the definition of (z, χ) (see (III.61)), we deduce:

$$\sum_{\alpha=1,2} \left\| \theta_\alpha - \theta_\alpha^h \right\|_{0,\omega} = \sup_{\varphi \in (L^2(\omega))^2} \left\{ \frac{k\left(z, \theta - \theta^h \right) + b\left(\chi, \theta - \theta^h \right)}{\sum_{\alpha=1,2} \| \varphi_\alpha \|_{0,\omega}} \right\}$$

or else, from the definition of θ, θ^h, χ and χ^h at (III.52) and (III.54):

$$\sum_{\alpha=1,2} \left\| \theta_\alpha - \theta_\alpha^h \right\|_{0,\omega} \le \sup_{\varphi \in (L^2(\omega))^2} \left\{ \frac{k\left(z - z^h, \theta - \theta^h \right) + b\left(\chi - \chi^h, \theta - \theta^h \right) + g\left(z^h \right) - g^h\left(z^h \right)}{\sum_{\alpha=1,2} \| \varphi_\alpha \|_{0,\omega}} \right\} .$$

But we proved that (see (III.63)):

$$\left\| z^h \right\|_{W_t} \le c \left[\sum_{\alpha=1,2} \| \varphi_\alpha \|_{0,\omega} \right]$$

and because of Theorem III.8:

$$\left| g\left(z^h \right) - g^h\left(z^h \right) \right| \le c\, h^2 \left(\sum_{\alpha=1,2} \| \varphi_\alpha \|_{0,\omega} \right) .$$

Finally:

$$\sum_{\alpha=1,2} \left\| \theta_\alpha - \theta_\alpha^h \right\|_{0,\omega} \le c \left[\sup_{\varphi \in (L^2(\omega))^2} \frac{\left\{ \| z - z^h \|_{W_t} + \| \chi - \chi^h \|_{M^h} \right\}}{\sum_{\alpha=1,2} \| \varphi_\alpha \|_{0,\omega}} \left\| \theta - \theta^h \right\|_{W_t} + h^2 \right]$$

and from (III.64), we deduce Theorem III.10.   ■

III.5.2   *Theoretical estimates on $u_3 - u_3^h$*

Let us recall that u_3 has been characterized in Chapter II as the solution to (see (II.69)):

(III.65)
$$\begin{cases} \qquad\quad \text{find } u_3 \in V_3, \text{ such that :} \\ \forall \ v_3 \in V_3, \ \int_\omega \partial_\alpha u_3 \, \partial_\alpha v_3 = -\int_\omega \theta_\alpha \, \partial_\alpha v_3 \end{cases}$$

where:

$$V_3 = \left\{ v \,|\, v \in H^1(\omega), v = 0 \text{ on } \gamma_0 \cup \gamma_1 \right\}.$$

The approximation of this problem is defined as follows:

(III.66)
$$\begin{cases} \text{find } u_3^h \in V_3^h, \text{ such that :} \\ \\ \forall \ v_3 \in V_3^h, \ \int_\omega \partial_\alpha u_3^h \, \partial_\alpha v_3 = -\int_\omega \theta_\alpha^h \, \partial_\alpha v_3 \end{cases}$$

where V_3^h is chosen for instance exactly as we did for V^h in the approximation of φ (see section III.5). Then one has the error estimate:

Theorem III.11
Assuming that hypothesis $\mathbf{H_1}$ *is valid, there exists a constant* c *such that*:

$$\| u_3 - u_3^h \|_{1,\omega} \le c \left[h^2 + \sup_{\varphi \in (L^2(\omega))^2} \frac{\left\{ \inf_{(\mu,\chi)} \| z - \mu \|_{w_t} + \| \chi - \Xi \|_M \right\}}{\sum_{\alpha = 1,2} \| \varphi_\alpha \|_{0,\omega}} + \inf_{v \in V_3^h} \| u_3 - v \|_{1,\omega} \right]$$

■

Proof of Theorem III.11
Let us first notice that for any element of the space and because of Poincaré inequality: for any $v \in V_3^h$:

$$c_1 \left\| u_3 - u_3^h \right\|_{1,\omega}^2 \le \int_\omega \left| \text{grad} \left(u_3 - u_3^h \right) \right|^2$$

$$= \int_\omega \partial_\alpha \left(u_3 - u_3^h \right) \partial_\alpha \left(v - u_3^h \right) + \int_\omega \partial_\alpha \left(u_3 - u_3^h \right) \partial_\alpha \left(u_3 - v \right)$$

$$\le \int_\omega \left(\theta_\alpha^h - \theta_\alpha \right) \partial_\alpha \left(v - u_3^h \right) + c_2 \left\| u_3 - u_3^h \right\|_{1,\omega} \left\| u_3 - v \right\|_{1,\omega}$$

and therefore, using Cauchy-Schwarz inequality:

$$\left\| u_3 - u_3^h \right\|_{1,\omega} \le c_3 \left\{ \sum_{\alpha = 1,2} \left\| \theta_\alpha - \theta_\alpha^h \right\|_{0,\omega} + \left\| u_3 - v \right\|_{1,\omega} \right\}$$

The proof of Theorem III.11 is then deduced from Theorem III.10.   ∎

It is also possible to improve the above estimates in the $L^2(\omega)$ norm. The result is contained in the following Theorem.

Theorem III.12
We assume that hypothesis $\mathbf{H_1}$ *is valid. Furthermore we assume that for any function* f_3 *in* $L^2(\omega)$ *the solution to (III.49) is in the space* $H^2(\omega)$. *Then there exists a constant* c *such that*

$$\left\| u_3 - u_3^h \right\|_{0,\omega} \le c \left[h^2 \left| u_3 \right|_{2,\omega} + \sum_{\alpha = 1,2} \left\| \theta_\alpha - \theta_\alpha^h \right\|_{0,\omega} \right]$$

                                                                                    ∎

Proof of Theorem III.12
From:

$$\left\| u_3 - u_3^h \right\|_{0,\omega} = \sup_{\varphi \in L^2(\omega)} \frac{\int_\omega \left(u_3 - u_3^h \right) \varphi}{\left\| \varphi \right\|_{0,\omega}}$$

and setting:

(III.67) $$\begin{cases} z \in V_3 \text{ such that:} \\ \\ \forall \ v \in V_3, \ \int_\omega \partial_\alpha z \, \partial_\alpha v = \int_\omega \varphi \, v \end{cases}$$

and

$$(III.68) \quad \begin{cases} z^h \in V_3^h \text{ such that:} \\ \\ \forall \ v \in V_3^h, \ \int_\omega \partial_\alpha z^h \partial_\alpha v = \int_\omega \varphi v \end{cases}$$

we deduce:

$$\|u_3 - u_3^h\|_{0,\omega} = \sup_{\varphi \in L^2(\omega)} \frac{\int_\omega \partial_\alpha z \partial_\alpha (u_3 - u_3^h)}{\|\varphi\|_{0,\omega}} \quad \text{and} \quad \|z^h\|_{1,\omega} \le c \|\varphi\|_{0,\omega}$$

or else, using equations (III.65) and (III.66):

$$(III.69) \quad \|u_3 - u_3^h\|_{0,\omega} \le \sup_{\varphi \in L^2(\omega)} \frac{\int_\omega \partial_\alpha (z - z^h) \partial_\alpha (u_3 - u_3^h) + \int_\omega (\theta_\alpha^h - \theta_\alpha) \partial_\alpha z^h}{\|\varphi\|_{0,\omega}}$$

But from (III.68), setting $v = z^h$, we obtain the *a priori* estimate:

$$\|z^h\|_{1,\omega} \le c \|\varphi\|_{0,\omega} .$$

Furthermore, the difference between (III.67) and (III.68) leads to the classical *a priori* error estimate:

$$\|z - z^h\|_{1,\omega} \le c \inf_{v \in V_3^h} \|v - z\|_{1,\omega} .$$

Then (III.69) enables one to derive the following inequality:

$$\|u_3 - u_3^h\|_{0,\omega} \le c_1 \sup_{\varphi \in L^2(\omega)} \left\{ \frac{\inf_{v \in V_3^h} \|v - z\|_{1,\omega}}{\|\varphi\|_{0,\omega}} \right\} \|u_3 - u_3^h\|_{1,\omega} + c_2 \sum_{\alpha=1,2} \|\theta_\alpha^h - \theta_\alpha\|_{0,\omega} . .$$

The proof is a consequence of the above estimate, and inequality (III.51). Let us notice that the validity of Theorem III.12 requires that u_3 solution to (III.65) is in the space $H^2(\omega)$. Otherwise the estimate obtained is meaningless. ∎

III.5.3   A first choice of finite elements

Case of triangles

Let us now define by R_K a function defined on the element K which is piecewise linear on K_i (see Figure III.14) for: $i = 1, 2, 3$; zero at each vertex of K and equal to one at the center of gravity G_K of K. Furthermore R_K is continuous on K.

$K = K_1 \cup K_2 \cup K_3$

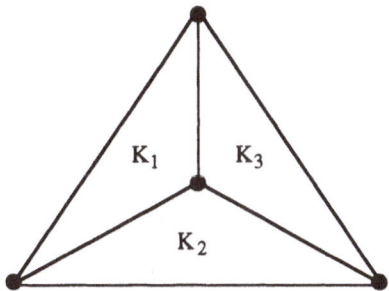

Figure III.14

Case of quadranles

For a quadrilateral we also consider vector functions (see Figure III.14) which are defined as follows:

On the reference element (one has three degrees of freedom on a quadrilateral):

$$(III.70) \qquad Z_K = A_1 \begin{pmatrix} B_K \\ 0 \end{pmatrix} + A_2 \begin{pmatrix} 0 \\ B_K \end{pmatrix} + A_3 R_K \begin{pmatrix} -\xi \\ \eta \end{pmatrix}$$

where A_1, A_2, A_3 are three constants, R_K is the piecewise bilinear function on each sub-quadrilateral K_i, zero on the vertices of K and equal to one at the center of gravity of K. Furthermore, R_K is continuous. The function B_K is the "bubble" on K, which is a "bi-quadratic function" vanishing at the vertices of K and equal to one at the center of gravity. As a matter of fact, B_K is the transformed function of a bi-quadratic function in a reference frame, by the bilinear mapping which maps a square in the reference frame onto the current quadrilateral. It is also a biquadratic function with respect to the physical coordinates for very special cases (the center of gravity of the current quadrilateral should be the intersection of the diagonals of the element).

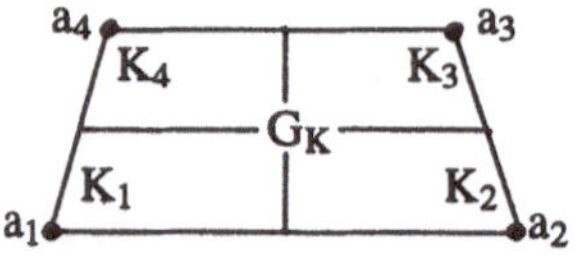

$$K = K_1 \cup K_2 \cup K_3 \cup K_4$$

Figure III.15

As a matter of fact, the vector functions Z_K generate a three-dimensional vector space of functions on K, vanishing on the boundary of this element. One can check immediately that the three vector functions involved in the definition of Z_K are linearly independent. The space generated by the vector functions Z_K is denoted by $\{Z_K\}$.

Let us now introduce the functional space used for the approximation of W_t and denoted by W_t^h. First of all, the mesh T^h is split into two parts. One is made of triangles and is denoted F^h. The other denoted by Q^h is made of quadrilaterals. Let us now introduce the approximation space for W_t by:

(III.71)
$$W_t^h = \left\{ \mu = (\mu_\alpha) \in W_t, \mu_{\alpha|K} \in P_1 \text{ if } K \in F^h \text{ and } \mu_{\alpha|K} \in Q_1 \text{ if } K \in Q^h \right\}$$
$$\cup \prod_{K \in F^h} \left\{ A_1 \begin{pmatrix} R_K \\ 0 \end{pmatrix} + A_2 \begin{pmatrix} 0 \\ R_K \end{pmatrix} \right\} \cup \prod_{K \in Q^h} \{Z_K\}$$

where P_1 is the set of first-degree polynomials and Q_1 the range of bilinear functions on a reference square by the mapping which transforms this square into the current element (hence the functions are basically homographic). The space M is by definition $L_0^2(\omega) \times \mathbb{R}^N$. It is approximated by:

(III.72) $M^h = \left\{ v \in L_0^2(\omega) \cap C^0(\overline{\omega}), v_{|K} \in P_1 \ \forall \ K \in F^h; v_{|K} \in Q_1 \ \forall \ K \in Q^h \right\} \times \mathbb{R}^N = H^h \times \mathbb{R}^N$

(P_1 and Q_1 being defined above).

III.5.3.1   *Interpolation of* $W_t \times M$ *by* $W_t^h \times M^h$

III.5.3.1.1 Interpolation of an element of M

Let us start with an element $\Lambda = (\psi , C)$ lying in the space M. Obviously, the element C of $\mathbb{R}^N$ is interpolated by itself. The function ψ is interpolated by the function $\pi\psi$ as follows. Let us denote by $\delta\psi^h$ the function piecewise linear on each element K of T^h (or bilinear which is equal to ψ at the nodes of the mesh). Then we add a constant to $\delta\psi^h$ such that $\pi\psi = \delta\psi^h + c \in L_0^2(\omega)$; i.e.:

$$c = \frac{-1}{\text{meas}(\omega)} \int_\omega \delta\psi^h .$$

From classical error estimates (see section III.1), we deduce that for any function $\psi \in L_0^2(\omega) \cap H^2(\omega)$:

$$\|\psi - \pi\psi\|_{0,\omega} \le \|\psi - \delta\psi^h\|_{0,\omega} + |c| \le c_1 h^2 |\psi|_{2,\omega} + |c| .$$

As $\psi \in L_0^2(\omega)$):

$$|c| = \left| \int_\omega \delta\psi^h \right| = \left| \int_\omega \delta\psi^h - \psi \right| \le \text{meas}(\omega)^{1/2} \|\delta\psi^h - \psi\|_{0,\omega} \le c_2 \text{meas}(\omega)^{1/2} h^2 |\psi|_{2,\omega}$$

Hence:

(III.73) $$\|\psi - \pi\psi\|_{0,\omega} \le c_3 h^2 |\psi|_{2,\omega}.$$

III.5.3.1.2 Interpolation of an element of the space W$_t$

Let us now consider an element θ of the space W_t. The interpolation of θ in the space W_t^h is defined as the unique element $\pi\theta \in W_t^h$ such that:

i) $\pi\theta = \theta$ at each node of the mesh,

ii) $\forall \psi \in H^h$, $\displaystyle\int_K \text{rot } \psi \cdot (\pi\theta - \theta) = 0$.

The second condition concerns mainly the internal degrees of freedom as we show in the following. It will facilitate the checking of hypothesis H_1. Let us make it explicit.

i) *case of a triangle*
If K is a triangle ψ is linear on K and thus:

$$\int_K \text{rot } \psi \cdot \left(\pi\theta - \theta\right) = \text{rot } \psi \int_K \left(\pi\theta - \theta\right) .$$

Hence condition ii) is equivalent to:

$$\forall \ \theta \in W_t , \ \int_K \left(\pi\theta - \theta\right) = 0$$

or else, because θ is a vector:

$$\int_K \pi\theta_1 = \int_K \theta_1 , \ \int_K \pi\theta_2 = \int_K \theta_2 .$$

These two relations can be easily satisfied by a correct choice of the internal degrees of freedom. This is always possible, because:

$$\int_K R_K = \frac{1}{3} \text{ meas }(K) > 0$$

where R_K is the Ramses function connected with the internal degrees of freedom.

ii)   *case of a quadrilateral*

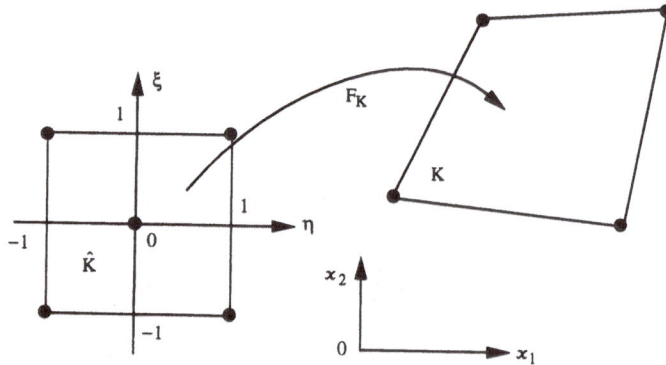

Figure III.16

If K is a quadrilateral, then ψ is the range of a bilinear function on a reference square – say $\widehat{K}$ – by the mapping which transforms K into $\widehat{K}$. Let us set:

$$\text{rot } \psi = \left(-\frac{\partial \psi}{\partial x_2}, \frac{\partial \psi}{\partial x_1} \right) = \left(-\frac{\partial \psi}{\partial \eta}\frac{\partial \eta}{\partial x_2} - \frac{\partial \psi}{\partial \xi}\frac{\partial \xi}{\partial x_2}, \frac{\partial \psi}{\partial \eta}\frac{\partial \eta}{\partial x_1} + \frac{\partial \psi}{\partial \xi}\frac{\partial \xi}{\partial x_1} \right)$$

where (η, ξ) are the coordinates in the reference frame (see Figure III.16). Let us set:

$$\begin{cases} \hat{\theta}_1 = |g| \left[\frac{\partial \xi}{\partial x_2} \theta_1 - \frac{\partial \xi}{\partial x_1} \theta_2 \right] \\ \hat{\theta}_2 = |g| \left[\frac{\partial \eta}{\partial x_1} \theta_2 - \frac{\partial \eta}{\partial x_2} \theta_1 \right] \end{cases}$$

where :

$$|g| = \frac{\partial x_1}{\partial \eta}\frac{\partial x_2}{\partial \xi} - \frac{\partial x_1}{\partial \xi}\frac{\partial x_2}{\partial \eta} \quad .$$

Then one has the equality:

($\widehat{\text{rot}} \, \psi$ is the rotational on the reference triangle $\hat{K}$: $\widehat{\text{rot}} \, \psi = \left(-\frac{\partial \psi}{\partial \eta}, \frac{\partial \psi}{\partial \xi} \right)$):

$$\int_K \text{rot } \psi \left(\pi\theta - \theta \right) = \int_{\hat{K}} \widehat{\text{rot}} \, \psi \left(\widehat{\pi\theta} - \hat{\theta} \right) \quad .$$

Furthermore, if we set $\psi = A + B\eta + C\xi + D\eta\xi$, then:

$$\widehat{\text{rot}} \, \psi = \left(-C - D\eta, B + D\xi \right)$$

and thus:

$$\int_{\hat{K}} \widehat{\text{rot}} \, \psi \left(\widehat{\pi\theta} - \hat{\theta} \right) = -C \int_{\hat{K}} \left(\widehat{\pi\theta} - \hat{\theta} \right)_1 + B \int_{\hat{K}} \left(\widehat{\pi\theta} - \hat{\theta} \right)_2 + D \int_{\hat{K}} \left[\xi \left(\widehat{\pi\theta} - \hat{\theta} \right)_2 - \eta \left(\widehat{\pi\theta} - \hat{\theta} \right)_1 \right] \quad .$$

Finally the condition (ii) is satisfied if and only if:

$$\begin{cases} \int_{\hat{K}} \left(\widehat{\pi\theta} - \hat{\theta} \right)_1 = 0 \ , \quad \int_{\hat{K}} \left(\widehat{\pi\theta} - \hat{\theta} \right)_2 = 0 \ , \\ \int_{\hat{K}} \left[\xi \left(\widehat{\pi\theta} - \hat{\theta} \right)_2 - \eta \left(\widehat{\pi\theta} - \hat{\theta} \right)_1 \right] = 0 \ . \end{cases}$$

These relations permit one to prescribe the values of the internal degrees of freedom of the element K. As a matter of fact, we just have to check that the last one is linearly independent of the first two ones. They define the coefficients of the bubble functions, B_K, with respect to the other degrees of freedom, and which involve separately the two components of a vector θ. The 3 x 3 matrix associated with the linear system above is:

$$
\begin{pmatrix}
\int_{\widehat{K}} B_{\widehat{K}} & 0 & -\int_{\widehat{K}} \xi\, R_{\widehat{K}} \\[2mm]
0 & \int_{\widehat{K}} B_{\widehat{K}} & \int_{\widehat{K}} \eta\, R_{\widehat{K}} \\[2mm]
-\int_{\widehat{K}} \eta\, B_{\widehat{K}} & \int_{\widehat{K}} \xi\, B_{\widehat{K}} & \int_{\widehat{K}} \left(\eta^2 + \xi^2\right) R_{\widehat{K}}
\end{pmatrix}
$$

($B_{\widehat{K}}$ being the bubble function and $R_{\widehat{K}}$ the Ramses one). A simple calculation leads to:

$$
\begin{pmatrix}
\dfrac{64}{9} & 0 & 0 \\[2mm]
0 & \dfrac{64}{9} & 0 \\[2mm]
0 & 0 & \dfrac{1}{3}
\end{pmatrix}
$$

the determinant of which being different from 0, the matrix is regular. Therefore it is always possible to choose the internal degrees of freedom in order to satisfy condition ii) for a quadrilateral as well. Finally one could easily check that π is continuous from $Wt \cap C^0(\overline{\omega})$ into W_t^h .

III.5.3.2   *Error estimates between the exact and approximate solution*

We are going to apply the general results obtained in sections III.5.1 and III.5.2. The first step consists in checking hypothesis H_1. Let us consider an element $\Lambda^h \in M^h$. We set $\Lambda^h = \left(\psi^h, c\right)$. Then:

$$
\forall\, \mu \in \left(H_0^1(\omega)\right)^2, \quad b\left(\Lambda^h, \mu\right) = \int_\omega \psi^h\, \text{curl}\, \mu = -\int_\omega \text{rot}\, \psi^h\, \mu
$$

and because of the properties of the interpolation $\pi\,\mu$ of μ (see section III.5.3.1):

$$\forall \; \mu \in \left(H_0^1 (\omega) \right)^2, \; b \left(\Lambda^h , \mu \right) = \int_\omega \psi^h \; \text{curl} \; \mu = - \int_\omega \text{rot} \; \psi^h \; \pi \mu = b \left(\Lambda^h , \pi \mu \right).$$

Hence : $\forall \; \mu \in \left(H_0^1 (\omega) \right)^2$, setting $\mu^h = \pi \mu$:

$$\sup_{\mu^h \in W_t^h} \frac{b \left(\Lambda^h , \mu^h \right)}{\| \mu^h \|_{W_t}} \geq \frac{b \left(\Lambda^h , \pi \mu \right)}{\| \pi \mu \|_{W_t}} \geq c \; \frac{b \left(\Lambda^h , \mu \right)}{\| \mu \|_{W_t}}$$

(we used the inequality: $\| \pi \mu \|_{W_t} \leq \frac{1}{c} \| \mu \|_{W_t}$ which expresses the continuity of the interpolation operator π). Hence:

$$\sup_{\mu^h \in W_t^h} \frac{b \left(\Lambda^h , \mu^h \right)}{\| \mu^h \|_{W_t}} \geq c \; \sup_{\mu \in (H_0^1(\omega))^2} \frac{b \left(\Lambda^h , \mu \right)}{\| \mu \|_{W_t}} \geq c \; \| \text{rot} \; \psi^h \|_{-1, \omega}$$

where $\| \text{rot} \; \psi^h \|_{-1, \omega} = \left\{ \| \partial_1 \psi^h \|_{-1, \omega}^2 + \| \partial_2 \psi^h \|_{-1, \omega}^2 \right\}^{1/2}$.

But from R. Temam [25] we know that there exists a constant such that:

$$\forall \; \psi \in L_0^2 (\omega), \; \| \psi \|_{0, \omega} \leq c_0 \; \| \text{rot} \; \psi \|_{-1, \omega} \; .$$

Hence we deduce that:

(III.74) $$\sup_{\mu^h \in W_t^h} \frac{b \left(\Lambda^h , \mu^h \right)}{\| \mu^h \|_{W_t}} \geq c_1 \; \| \psi^h \|_{0, \omega} \; .$$

Let us now consider an element μ^h in the space W_t^h such that μ^h is zero at all the nodes internal to ω, $\mu_\alpha^h \; b_\alpha = 0$ at all the nodes on the boundary of ω and $\mu_\alpha^h \; a_\alpha = c_i$ at the nodes of γ_2^i (except the ones at the extremities of γ_2^i which are in common with γ_0 and (or) γ_1). As μ^h is piecewise linear on the edges of the elements (even for quadrilaterals), $\mu_\alpha^h \; a_\alpha$ is constant on γ_2^i except – maybe – on the last vertices in the vicinity of γ_1 or γ_0. Therefore one has:

$$\int_{\gamma_2^i} \mu_\alpha^h \; a_\alpha = C_i \; K_i \qquad K_i \geq K_0 \quad \forall \; i = 1 , N$$

where N is the number of connected components of the free edge and K_0 is a constant (strictly positive). But one has also for this particular choice of μ^h:

$$\sup_{v \in W_t^h} \frac{b\left(\Lambda^h, v\right)}{\|v\|_{W_t}} \geq \frac{b\left(\Lambda^h, \mu^h\right)}{\|\mu^h\|_{W_t}} \geq \sum_{i=1, N_2} \frac{|C_i|^2 K_0}{\|\mu^h\|_{W_t}} - M_1 \|\psi^h\|_{0, \omega}$$

where we used the following estimate:

$$\left| \int_\omega \psi^h \ \mathrm{curl} \ \mu^h \right| \leq M_1 \|\psi^h\|_{0, \omega} \|\mu^h\|_{W_t}$$

and because of the continuity of the interpolation operator π from W_t into W_t^h (details are explicited in step 2 section *III.5.4.1*):

$$\|\mu^h\|_{W_t} \leq M_2 \left[\sum_{i=1, N_2} |C_i|^2 \right]^{1/2}$$

then:

(III.75) $$\sup_{v \in W_t^h} \frac{b\left(\Lambda^h, v\right)}{\|v\|_{W_t}} \geq \frac{K_0}{M_2} \left[\sum_{i=1, N_2} |C_i|^2 \right]^{1/2} - M_1 \|\psi^h\|_{0, \omega} \ .$$

Finally, combining (III.74) and (III.75), we deduce the validity of hypothesis **H₁**.

Let us now establish the main result of this section, which concerns the error between the exact and approximate solutions.

Theorem III.13
Let us assume that the plate model solution is such that:

$$u_3 \in H^3(\omega) \left(\text{hence: } \theta_\alpha \in H^2(\omega), \ \psi \in H^1(\omega)\right)$$

Let us also assume that the solution (z, χ) to III.61 is such that (we set $\chi = (\lambda, C)$):

$$\|z\|_{2, \omega} + \|\lambda\|_{1, \omega} \leq \sum_{\alpha=1, 2} \|\varphi_\alpha\|_{0, \omega} \quad \text{where}: \|z\|_{2, \omega} = \sum_{\alpha=1, 2} \|z_\alpha\|_{2, \omega}$$

φ being the fonction at the right hand side of (III.61).
Furthermore, we assume that $\varphi \in H^2(\omega)$, φ being that time the solution to (III.49). This is an independent assumption from the previous one because φ is determined separately. Then there exists a positive constant c such that:

$$① \; \left\| u_3 - u_3^h \right\|_{1,\omega} \leq c\,h \qquad , \; ② \; \left\| \varphi - \varphi^h \right\|_{1,\omega} \leq c\,h$$

$$③ \; \left\| \theta_\alpha - \theta_\alpha^h \right\|_{1,\omega} \leq c\,h \qquad , \; ④ \; \left\| \psi - \psi^h \right\|_{0,\omega} \leq c\,h$$

$$⑤ \; \left\| u_3 - u_3^h \right\|_{0,\omega} \leq c\,h^2 \qquad , \; ⑥ \; \left\| \varphi - \varphi^h \right\|_{0,\omega} \leq c\,h^2$$

$$⑦ \; \left\| \theta_\alpha - \theta_\alpha^h \right\|_{0,\omega} \leq c\,h^2$$

∎

Proof of Theorem III.13

Estimates ② and ⑥ are very classical and have already been mentioned in (III.51). The estimates ③ and ④ are deduced from Theorem III.9 by choosing elements μ and Ξ, which are classical Lagrange interpolations of θ and ψ (it is not necessary to use the internal degree of freedom of W_t; therefore error estimates are directly derived from those of Raviart-Thomas [3] or Ciarlet [2], the "inf" being obviously smaller than the value obtained with the interpolation). The $L^2(\omega)$ error estimates ⑦ are a consequence of Theorem III.10. As a matter of fact, one has:

$$\inf_{(\mu,\Xi) \,\in\, W_t^h \times M^h} \left\{ \left\| z - \mu \right\|_{W_t} + \left\| \chi - \Xi \right\|_M \right\} \leq c\,h \left[|z|_{2,\omega} + |\chi|_{1,\omega} \right]$$

where we set $\Xi = (\lambda, c) \in L_0^2(\omega) \times \mathbb{R}^N$ and we admit that (III.61) satisfies the regularity assumption mentioned in the Theorem:

$$(\text{III.76}) \qquad \left[\|z\|_{2,\omega} + \|\lambda\|_{1,\omega} \right] \leq c \sum_{\alpha=1,2} \|\varphi_\alpha\|_{0,\omega} \; .$$

Hence ⑦ is directly deduced from Theorem III.10. Concerning inequalities ① and ⑤ the proof is deduced from Theorems III.11 and III.12 (assuming again that (III.76) is true). ∎

III.5.4   A second choice of finite elements

Concerning the approximation of elements Λ in the space $M = L_0^2(\omega) \times \mathbb{R}^N$, the same space M^h as the one used in the first choice in section III.5.3. is adopted. The difference lies in the definition of W_t^h. Hence we set:

$$(\text{III.77}) \qquad \begin{aligned} W_t^h = \big\{ &\mu = (\mu_\alpha) \in W_t \,, \; \mu_\alpha \in C^0(\overline{\omega}) \,, \\ &\mu_{\alpha|K} \in P_2 \; \text{if} \; K \in F^h \; \text{and} \; \mu_{\alpha|K} \in Q_2 \; \text{if} \; K \in Q^h \big\} \end{aligned}$$

or if we add the internal degrees of freedom:

$$(\text{III.78}) \quad W_t^h = \left\{ \mu = (\mu_\alpha) \in W_t, \ \mu_{\alpha|K} \in P_2 \text{ if } K \in F^h \text{ and } \mu_{\alpha|K} \in Q_2 \text{ if } K \in Q^h \right\}$$

$$\cup \prod_{K \in F^h} \left\{ A_1 \begin{pmatrix} R_K \\ 0 \end{pmatrix} + A_2 \begin{pmatrix} 0 \\ R_K \end{pmatrix} \right\} \cup \prod_{K \in Q^h} \{ Z_K \} \quad .$$

The first step in the error analysis is to check hypothesis H_1. Obviously, if it is satisfied for the definition (III.77) of W_t^h, it is also true for (III.78) because of the space embedding. Furthermore, the definition (III.71) led to a subspace of the one defined in (III.78) As hypothesis H_1 is satisfied for the space W_t^h introduced in (III.71), it is automatically satisfied with definition (III.78). Finally, the only case to be checked is the definition (III.77) of W_t^h.

Triangle

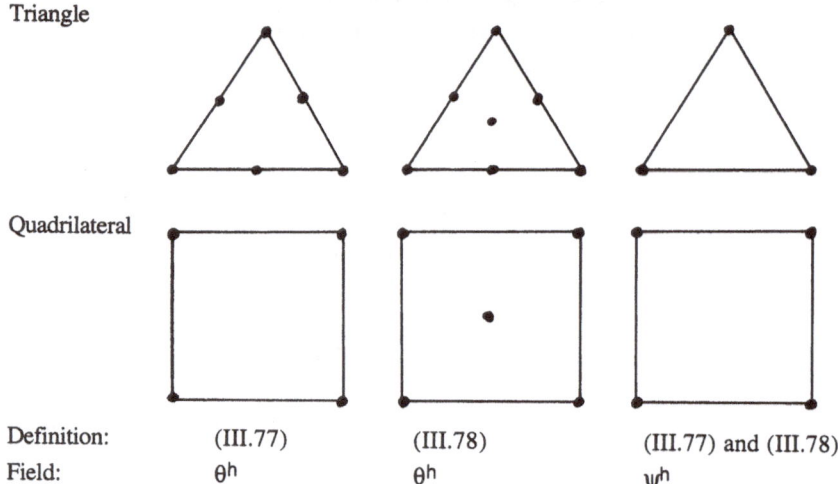

Quadrilateral

Definition:	(III.77)	(III.78)	(III.77) and (III.78)
Field:	θ^h	θ^h	ψ^h

Figure III.17

III.5.4.1   Checking hypothesis H_1 with the definition (III.77) of W_t^h ((III.78) being obvious)

There are three steps in the proof.

<u>*Step 1*</u>

Let us redefine the approximation space for ψ by:

$$H^h = \left\{ \psi \mid \psi \in L_0^2(\omega) \cap C^0(\overline{\omega}), \ \forall \ K \in F^h, \ \psi_{|K} \in P_1 \ ; \ \forall K \in Q^h, \ \psi_{|K} \in Q_1 \right\} \quad .$$

Let us also recall that F^h is a collection of triangles and Q^h a collection of quadrilaterals. Let us now consider an element θ of the space W_t^h which is zero on the boundary of ω (always assumed to be piecewise linear). Then one has:

$$\forall\ \psi \in H^h\ ,\ \int_\omega \psi\ \text{curl}\ \theta = -\int_\omega \text{rot}\ \psi \cdot \theta\ .$$

Let a_i denote a mid-point of an edge in the mesh T^h (i.e. a node which is different from a vertex; see Figure III.18). If θ is a vector function in the space W_t^h vanishing on the boundary of ω and also being zero at all the nodes of the mesh except at point a_i (see Figure III.18), one has:

$$\int_\omega \text{rot}\ \psi \cdot \theta = \int_{K_i^1} \text{rot}\ \psi \cdot \theta + \int_{K_i^2} \text{rot}\ \psi \cdot \theta$$

Figure III.18

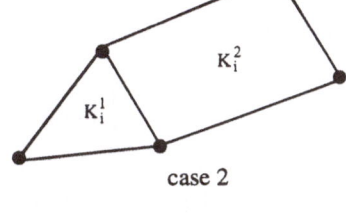

case 2

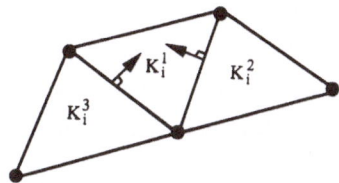

case 1

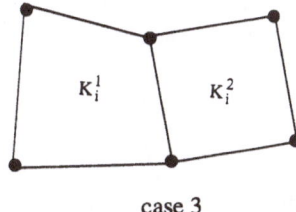

case 3

Figure III.19

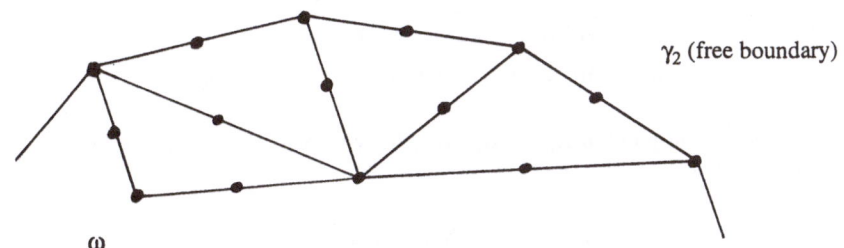

γ_2 (free boundary)

ω

The two elements K_i^1 and K_i^2 are adjacent to the node a_j. The function ψ is continuous at the interface between K_i^1 and K_i^2. Hence, if b^1 is the unit normal (the components are b_α^1 for $\alpha = 1,2$), to the common edge between K_i^1 and K_i^2, one has:

$$\left[\frac{\partial \psi}{\partial s}\right] = (\text{rot } \psi, b^1) = 0$$

where [.] denotes the jump of a quantity along $K_i^1 \cap K_i^2$. Let us then set:

$$\begin{cases} \theta^1 \in W_t^h, \ \theta^1 = 0 \text{ on } \partial\omega \\ \theta^1 (a_i) = (\text{rot } \psi, b^1) b^1 \\ \theta^1 = 0 \text{ at each node of the mesh} \neq a_i \end{cases}$$

One has the following relationship, if K_i^1 and K_i^2 are triangles :

$$\int_\omega (\text{rot } \psi, \theta^1) = \frac{(\text{rot } \psi, b^1)^2}{3} \left[\text{mes}\left(K_i^1\right) + \text{mes}\left(K_i^2\right)\right]$$

We used the integration formula based on the three mid-points of a triangle. This is possible because the shape functions of W_t^h are P_2. In the following, we focus on the case where K_i^1 and K_i^2 are triangles. Quadrilaterals could be treated in a very similar way.

Let us consider triangle K_i^1 for instance. There exists another vertex of K_i^1 with a normal b^2 (see Figure III.18) such that:

$$\int_\omega (\text{rot } \psi, \theta^2) = \frac{(\text{rot } \psi, b^2)^2}{3} \left[\text{mes}\left(K_i^1\right) + \text{mes}\left(K_i^3\right)\right]$$

with

$$\begin{cases} \theta^2 \in W_t^h, \ \theta^2 = 0 \text{ on } \partial\omega \\ \theta^2 (a_k) = (\text{rot } \psi \cdot b^2) b^2 \\ \theta^2 = 0 \text{ at each node of the mesh} \neq a_k \end{cases}$$

a_k being the mid-point of $K_i^1 \cap K_i^3$. But the two vectors b^1 and b^2 are linearly independent. Hence one has:

$$|\text{rot } \psi|^2 \leq c \left[(\text{rot } \psi, b^1)^2 + (\text{rot } \psi, b^2)^2\right]$$

where c is a constant which does not depend on the mesh size h, but only on the angle between b^1 and b^2 because it is just a change of basis. Hence we assumed that this angle is bounded from below and above in order to fix c. Finally we proved that there exists a vector field θ_{K_i} in W_t^h, vanishing on the boundary of ω and such that:

(III.79)
$$\int_\omega \psi \, \mathrm{curl} \; \theta_{K_i} \geq c_1 \, |\mathrm{rot} \; \psi|^2_{K_i} \, \mathrm{mes} \, (K_i)$$

with

$$\theta_{K_i} = \theta^1 + \theta^2 \; .$$

Then one has, using the inverse inequality (see Ciarlet [2] for instance), for a basis function on K (or a direct calculation):

$$\|\theta_{K_i}\|^2_{W_t} \leq \frac{c}{h} \Big[|\mathrm{rot} \; \psi|^2_{K_i^1} \, \mathrm{mes} \, (K_i^1) + |\mathrm{rot} \; \psi|^2_{K_i^2} \, \mathrm{mes} \, (K_i^2) \Big]$$

and if we set:

$$\theta^0 = \sum_{K \in T^h} \theta_K$$

then:

(III.80)
$$\|\theta^0\|_{W_t} \leq \frac{c}{h} \, |\psi|_{1,\omega}$$

where c is a positive constant which is h-independent. But from (III.79)

(III.81)
$$\int_\omega \psi \, \mathrm{curl} \; \theta^0 = \sum_{K \in T^h} \int_K \psi \, \mathrm{curl} \; \theta_K \geq c \, |\psi|^2_{1,\omega}$$

Let us notice that the constant c appearing in the above and below expressions are distinct. Finally from (III.80) and (III.81):

(III.82)
$$\sup_{\theta \in W_t^h} \frac{\displaystyle\int_\omega \psi \, \mathrm{curl} \; \theta}{\|\theta^0\|_{W_t}} \geq c \, h \, |\psi|_{1,\omega} \; .$$

Step 2

Let us now notice the following inequality which is valid for any element ψ in the space H^h:

$$\sup_{\theta \in W_t^h} \frac{\int_\omega \psi \, \text{curl} \, \theta + \sum_{i=1,N} C_i \int_{\gamma_2^i} \theta_s}{\|\theta\|_{W_t}} \geq \sup_{\theta \in W_t^h} \frac{\sum_{i=1,N} C_i \int_{\gamma_2^i} \theta_s}{\|\theta\|_{W_t}} - K \, \|\psi\|_{0,\omega}$$

K being a positive constant independent of the mesh size h. There are several methods for minimizing the first term on the right-hand side of this relation. Let us give one of them. First of all, it is sufficient to consider one term. Then, one has:

(III.83)
$$\sup_{\theta \in W_t^h} \frac{C_i \int_{\gamma_2^i} \theta_s}{\|\theta\|_{W_t}} \geq K_1 |C_i| .$$

This is obtained in two steps. First of all, we choose θ as an element of the space W_t such that:

$$\int_{\gamma_2^i} \theta_s = \text{sign} \, (C_i)$$

and the prolongation of θ inside ω being continuous from $H_{00}^{1/2} \left(\gamma_2^i \right)$ into W_t (there exists at least one! see J.L. Lions - E. Magenes [7]), one has the estimate (III.83). Let us now consider the interpolation of θ, the one used for the estimate (III.83), which is denoted by $\pi\theta$. Thus:

$$\sup_{\theta \in W_t^h} \frac{C_i \int_{\gamma_2^i} \theta_s}{\|\theta\|_{W_t}} \geq K_1 |C_i| - |C_i| \frac{\|\theta_s - (\pi\theta)_s\|_{1/2, \gamma_2^i}}{\|\theta\|_{W_t}} .$$

But from the continuity of the trace operator from W_t (i.e. $H^1 (\omega)$) into $H^{1/2} \left(\gamma_2^i \right)$) one has:

$$\|\theta_s - (\pi\theta)_s\|_{1/2, \gamma_2^i} \leq c \, \|\theta - \pi\theta\|_{W_t}$$

The vector field θ which satisfies the condition (III.83) can be assumed to be $H^2 (\omega)$ for each of its components. This is possible because among all the functions satisfying the condition:

$$\int_{\gamma_2^i} \theta_s = \text{sign} \, (C_i)$$

it is always possible to choose one which is smooth enough. Hence there is a constant – say K_2 – independent of h and such that:

$$\| \theta_s - (\pi \theta)_s \|_{1/2, \gamma_2^i} \leq K_2 \, h^2$$

(see the finite element error estimates from Raviart-Thomas [3] or Ciarlet [2]). Finally, we proved that for h small enough, one has:

$$\sup_{\theta \in W_t^h} \frac{C_i \displaystyle\int_{\gamma_2^i} \theta_s}{\| \theta \|_{W_t}} \geq K_3 |C_i| \ .$$

Hence there exists a constant K_4, independent of the mesh size, with the classical assumptions on the regularity of the mesh family, such that:

$$(\text{III.84}) \qquad \sup_{\theta \in W_t^h} \frac{\left\{ \displaystyle\int_\omega \psi \, \mathrm{curl}\, \theta + \sum_{i=1,N} C_i \int_{\gamma_2^i} \theta_s \right\}}{\| \theta \|_{W_t}} \geq K_4 \left[\sum_{i=1,N} |C_i| - \| \psi \|_{0,\omega} \right]$$

and with (III.82) on the one hand and the inverseinequality:

$$\left(\exists \, c > 0, \ \forall \, \psi \in H^h \, | \psi |_{1,\omega} \leq \frac{c}{h} \| \psi \|_{0,\omega} \right)$$

on the other hand:

$$(\text{III.85}) \ \sup_{\theta \in W_t^h} \frac{\left\{ \displaystyle\int_\omega \psi \, \mathrm{curl}\, \theta + \sum_{i=1,N} C_i \int_{\gamma_2^i} \theta_s \right\}}{\| \theta \|_{W_t}} \geq K_5 \left[\sum_{i=1,N} |C_i| + h |\psi|_{1,\omega} \right]$$

($| \ |_{1,\omega}$ is a norm on the space $H^1(\omega) \cap L_0^2(\omega)$ which is equivalent to the one induced by the space $H^1(\omega)$).

Step 3

Let us now show how it is possible to replace the term $h \, |\phi|_{1,\omega}$ in the above estimate by $\| \psi \|_{0,\omega}$. First of all we set:

$$W_t^0 = \left(H_0^1(\omega) \right)^2$$

and we define for each function ψ in H^h the solution z to the following system:

$$\left|\begin{array}{l} z \in W_t^0 \text{ such that:} \\[2mm] \forall \mu \in W_t^0 , \ k(z,\mu) = - \int_\omega \psi \operatorname{curl} \mu \end{array}\right.$$

where $k(.,.)$ is the bilinear form on W_t which has been defined earlier. Then we associate the approximate solution z^h (of z) by:

$$\left|\begin{array}{l} z^h \in W_t^0 \cap W_t^h \text{ such that:} \\[2mm] \forall \mu \in W_t^0 \cap W_t^h , \ k(z^h,\mu) = - \int_\omega \psi \operatorname{curl} \mu \end{array}\right.$$

Classical error estimates, assuming that the involved operator is smooth enough, lead to the inequality:

$$\| z - z^h \|_{W_t^0} \leq c \, h \sum_{\alpha=1,2} |z_\alpha|_{2,\omega}$$

with $z = (z1 , z2)$. But because of the regularity assumption mentioned above (the one of the operator associated to the bilinear form $k(.,.)$), we have also:

$$\sum_{\alpha=1,2} |z_\alpha|_{2,\omega} \leq K_6 |\psi|_{1,\omega}$$

Let us notice that $\forall \ \mu \in W_t^0$, $- \int_\omega \psi \operatorname{curl} \mu = \int_\omega \operatorname{rot} \psi \cdot \mu)$. Thus:

$$\| z - z^h \|_{W_t} \leq c \, K_6 \, h^2 \, |\psi|_{1,\omega}$$

Let us now consider the projection μ^h of an element μ of W_t^0 onto W_t^h such that:

$$\left|\begin{array}{l} \mu^h \in W_t^h \cap W_t^0 \\[2mm] \forall \ q \in W_t^h \cap W_t^0 , \ k(q,\mu-\mu^h) = 0 \end{array}\right.$$

Then one has for any μ in the space W_t^0 :

$$-\int_\omega \psi \, \mathrm{curl} \, \mu = -\int_\omega \psi \, \mathrm{curl} \left(\mu - \mu^h\right) - \int_\omega \psi \, \mathrm{curl} \, \mu^h$$

$$= k\left(z, \mu - \mu^h\right) - \int_\omega \psi \, \mathrm{curl} \, \mu^h = k\left(z - z^h, \mu - \mu^h\right) - \int_\omega \psi \, \mathrm{curl} \, \mu^h$$

$$= k\left(z - z^h, \mu\right) - \int_\omega \psi \, \mathrm{curl} \, \mu^h$$

$$\leq K_7 \left\|z - z^h\right\|_{W_t} \cdot \|\mu\|_{W_t} - \int_\omega \psi \, \mathrm{curl} \, \mu^h$$

or else:

(III.86) $$\qquad -\int_\omega \psi \, \mathrm{curl} \, \mu \leq K_8 \, h \, |\psi|_{1,\omega} \cdot \|\mu\|_{W_t} - \int_\omega \psi \, \mathrm{curl} \, \mu^h \, .$$

Let us notice that from the definition of μ^h (as the projection on $W_t^0 \cap W_t^h$ of μ):

$$K_9 \left\|\mu^h\right\|_{W_t}^2 \leq k\left(\mu^h, \mu^h\right) = k\left(\mu, \mu^h\right) \leq K_{10} \|\mu\|_{W_t} \left\|\mu^h\right\|_{W_t}$$

which implies:

$$\left\|\mu^h\right\|_{W_t} \leq K_{11} \|\mu\|_{W_t} \qquad \left(\text{where } K_{11} = \frac{K_{10}}{K_9}\right)$$

and with (III.86):

$$\forall \, \mu \in W_t^0, \quad \frac{-\displaystyle\int_\omega \psi \, \mathrm{curl} \, \mu}{\|\mu\|_{W_t}} \leq K_8 \, h \, |\psi|_{1,\omega} + K_{11} \frac{\left|\displaystyle\int_\omega \psi \, \mathrm{curl} \, \mu^h\right|}{\left\|\mu^h\right\|_{W_t}}$$

and, because $\psi \in L_0^2(\omega)$ (see R. Temam [25] for instance):

$$\sup_{\mu \, \in \, W_t^0} \frac{\displaystyle\int_\omega \psi \, \mathrm{curl} \, \mu}{\|\mu\|_{W_t}} = \|\mathrm{rot} \, \psi\|_{-1,\omega} \geq K_{12} \|\psi\|_{0,\omega}$$

($\mathrm{rot} \, \psi$ is a vector and $\|\mathrm{rot} \, \psi\|_{-1,\omega}$ is the norm $H^{-1}(\omega)$ of the components), we deduce that:

(III.87)
$$\sup_{\mu \in W_t^h} \frac{\int_\omega \psi \, \mathrm{curl}\, \mu}{\|\mu\|_{w_t}} \geq \sup_{\mu \in W_t^0 \cap W_t^h} \frac{\int_\omega \psi \, \mathrm{curl}\, \mu}{\|\mu\|_{w_t}}$$

$$\geq \frac{K_{12}}{K_{11}} \|\psi\|_{0,\omega} - \frac{K_8}{K_{11}} h \, |\psi|_{1,\omega} .$$

Let us now summarize estimates (III.82), (III.85) and (III.87). By a simple linear combination we obtain the final result of this section:

(III.88) $$\sup_{\mu \in W_t^h} \left\{ \left| \frac{\int_\omega \psi \, \mathrm{curl}\, \mu}{\|\mu\|_{w_t}} + \sum_{i=1,N} \frac{C_i \int_{\gamma_2} \mu_s}{\|\mu\|_{w_t}} \right| \right\} \geq K_{13} \left\{ \|\psi\|_{0,\omega} + \sum_{i=1,N} |C_i| \right\}$$

which establishes the validity of hypothesis $\mathbf{H_1}$ (needed in the error estimates).   ■

Comments on the proof given in this section

For the Stokes problem, Glowinski and Pironneau [26] introduced a mathematical analysis of the so-called Taylor-Hood element [27]. The formulation is very similar to the one used in this section. But the potential ψ plays the role of the pressure and the vector θ is assimilated to the velocity field up to a rotation of angle $\frac{\pi}{2}$. The "curl" operator is therefore transformed into the "div" (for Stokes). Then the elements used in this section (P_2 or Q_2 for θ and P_1 or Q_1 for ψ) are equivalent to the Taylor-Hood element [27]. The new difficulty is mainly contained in the boundary conditions (in the mathematical studies of Stokes problem the boundary conditions are usually equivalent to $\theta_\alpha = 0$ on all the boundary). Let us point out that it would be unrealistic to limitate the analysis to such a situation. Furthermore an important difficulty arise due to multiconnected boundary case.

For the proof of hypothesis $\mathbf{H_1}$, we used the idea of Bercovier-Pironneau [28] which has been reproduced in Girault-Raviart [29]. But the method is slightly different because of the boundary conditions which can be quite arbitrary in our case. The mathematical trick (Step 3) that is presented here was formerly introduced in Destuynder-Nedelec [30] and then adapted to the finite elements studied in this section by Destuynder-Nevers [31] , [32]. The extension to shell could be done with additional technical difficulties in the writing of the proof. But it works perfectly. In chapter VIII, the analysis is limited to the first order element analyzed for plates in section III.5.3.

III.5.4.3   Error estimates

The error estimates concerning the potential φ are the same as the ones obtained at (III.5.1) and Theorem III.8, because φ is determined independently and the finite element approximation is the same in both cases. The first result concerns the error between θ and θ^h on the one hand and the error between Λ and Λ^h on the other hand.

Theorem III.14
Let us assume that the solution (θ , Λ) to the model (III.52) is such that:

$$\theta_\alpha \in H^3(\omega), \ \psi \in H^2(\omega)$$

and that the regularity assumptions needed to check Theorem III.9 *are satisfied. Then there exists a constant – say K – which is independent on the mesh size* h *and such that:*

$$\left\| \theta - \theta^h \right\|_{W_t} + \left\| \Lambda - \Lambda^h \right\|_{M^h} \leq K h^2$$

(obviously K depends on θ and Λ in the norms $H^3(\omega)$ and $H^2(\omega) \times \mathbb{R}^N$ respectively).   ■

Proof of Theorem III.14
The proof comes straightforwardly from Theorem III.4 and the classical error estimates in finite element method (see Raviart-Thomas [3] or Ciarlet [2]).   ■

Remark III.8
The $L^2(\omega)$ error estimates cannot be used with any improvement in this case because the error estimate on the right-hand side of equation (III.52) characterizing (θ , ψ) is only approximated up to $0 (h^2)$. This limits the quality of the approximation of (θ , ψ).   ■

REFERENCES

[1] STRANG G., FIX G., [1973], An analysis of the finite element method, Series in Automatic computation, Prentice Hall, New York.

[2] CIARLET P.G., [1978], The finite element method for elliptic problems, Studies in Mathematics and its applications n° 4, North-Holland, Amsterdam.

[3] RAVIART P.A., THOMAS J.M., [1983], Introduction à l'analyse numérique des équations aux dérivées partielles, Masson, Paris.

[4] R.H. MACNEAL [1978], A simple quadrilateral shell element, Comput. Struc., 8, p 175-183.

[5] HUGHES J.R., TAYLOR R.L., KANOKNUKULCHAI W., [1977], A simple and efficient element for plate bending, Int. J. Numer. Meth. Engrg., 11, (10), p 1529-1543.

[6] HUET D., [1976], Décomposition spectrale et opérateurs, P.U.F., le mathématicien, Paris.

[7] LIONS J.L., MAGENES E., [1968], Problèmes aux limites non homogènes et applications vol 1, Dunod, Paris.

[8] ARGYRIS J.H., SCHARPF D.W., [1968], The sheba family of shell elements for the matrix displacement method, Aeron J.R. Aeron. Soc., 72, p 878-883.

[9] BERNADOU M., [1978], Thèse d'état (Université Pierre et Marie Curie, Laboratoire d'Analyse Numérique, Paris) and Convergence of conforming finite element methods for general shell problems, [1980], Internat. J. Engrg. Sci., 18, p 249-276.

[10] ARGYRIS J.H., FRIED J.,. SCHARPF D.W., [1968], The TUBA family of plate elements for the matrix displacement method - The Aeronautical Journal of the Royal Aeronautical Society, 72, p 701-709.

[11] PARISCH H., [1981], Large displacements of shells including material nonlinearities - Comp. Meth. Appl. Mech. Engrg, 27, p 183-214.

[12] IRONS B.M., RAZZAQUE A., [1972], Experience with the patch test for convergence of finite elements ; in the Mathematical Foundations of the finite element Method with Applications to Partial Differential Equations (A. K. Aziz editor), p 557-587.

[13] ZIENKIEWICZ O.C., [1977], The finite element method, McGraw-Hill, London, 3rd edition.

[14] LASCAUX P., LESAINT P., [1975], Some non-conforming finite elements for the plate

bending problem, RAIRO série Rouge Anal. Numer., R1, p 9-53.

[15] ARNOLD D., FALK R., [1990], The boundary layer for the Reissner-Mindlin plate model, Siam J. Math. Numer. Anal..

[16] ADINI A., CLOUGH R.W., [1961], Analysis of plate bending by the finite element method - NSF report G7337.

[17] GRISVARD P., [1985], Elliptic problems in non smooth domains, Pitman, Boston.

[18] LIONS J.L., [1973], Perturbations singulières dans les problèmes aux limites et en contrôle optimal, Lecture notes in mathematics, Springer-Verlag, Berlin.

[19] DESTUYNDER Ph., [1986], Une théorie asymptotique des plaques minces en élasticité linéaire, RMA2, Masson, Paris.

[20] FRIEDRICHS K.O., DRESSLER R.F., [1961], A boundary layer theory for elastic plates CPAM XIV, p 1-33.

[21] GOL'DENVEIZER A.L., [1962], Derivation of an approximate theory of bending a plate by the method of asymptotic integration of the equations of the theory of elasticity, PRIKL., MATH., MECH., 26, p 668-686 (Traduction PMM19A).

[22] BREZZI F., [1974], On the existence, uniqueness and approximation of saddle point problems arising from Lagrangian multipliers, RAIRO-R2 p 129-151.

[23] ADAMS R., [1976], Sobolev Spaces, Academic Press, New-York.

[24] BELYTSHKO T., STOLARSKI M., [1983], Shear and membrane locking in curved Co elements comput. Methods Appl. Mech. Eng., 41, p 279-296.

[25] TEMAM R., [1979], Navier Stokes equations, Studies in mathematics Vol. 2, North-Holland, Amsterdam.

[26] GLOWINSKI R., PIRONNEAU O., [1976], Sur la résolution numérique du problème de Dirichlet pour l'opérateur biharmonique par une méthode quasi-directe, C.R. Acad. Sci. Paris série A, 282, p 223-226.

[27] TAYLOR C., HOOD P., [1973], A numerical solution of the Navier Stokes equations using the finite element technique, Comp. and Fluids 1, p 73-100.

[28] BERCOVIER M., PIRONNEAU O., [1979], Error estimates for finite element method solution of the Stokes problem in the primitive variables, Numer. Math. 33, p 211-224.

[29] GIRAULT V., RAVIART P.A., [1986], Finite element methods for Navier-Stokes equations, Computational mathematics n° 5, Springer-Verlag, Berlin.

[30] DESTUYNDER Ph., NEDELEC J.C., [1986], Approximation numérique du cisaillement transverse dans les plaques minces en flexion. Numer. Math., 48, p 281-302.

[31] DESTUYNDER Ph., NEVERS Th., [1988], Some numerical aspects of mixed finite elements for bending plates comp. Meth. Appl. Mechs Eng., 78, p 73-87.

[32] DESTUYNDER Ph., NEVERS Th., [1986], A new finite element scheme for bending plates. Comput Meth. Appl. Sci. Eng., 68, p 127-139.

Chapter 4

NUMERICAL TESTS FOR THE MIXED FINITE ELEMENT SCHEMES FOR BENDING PLATES

IV.0   A brief description of the chapter

The numerical performances of the finite element schemes discussed in the previous chapters are presented. The results obtained are compared to the ones of the QUAD 4 element of MacNeal. Additionnally the solution methods are evaluated, and the vector and parallel optimization has been used in order to deliver the best computational time.

This chapter was written jointly with Thierry Nevers in 1988 and was published a few years later in "La Recherche Aérospatiale" which is published by ONERA.

IV.1   Precision tests for the mixed formulation

IV.1.1   A recall of the equations to be solved

Because of the similarity of the equations obtained with the Navier-Stokes equations, we tested the conventional finite elements used in fluid mechanics and which are well-known for their "good performance". The best known is the so-called Taylor-Hood element (without bubbles) which has been presented in section III.5.4 (definition (III.77) of W_t^h) These elements (used in the following tests) are summarized in table I and II.

Table I. Triangle type elements

Table II. Quadrangle type elements

Function	θ	ψ	u_3	φ
Element				

The error estimates are the those obtained at Theorem III.9: $O(h^2)$ for Θ in W_t norm and for Ψ in L_0^2 norm. Actually, we only have $o(h)$ for u_3 and Φ in H^1 (ω) norm, but this does not alter the precision on Θ and Ψ. Let us recall also that the transverse shear Q is approximated by:

$$Q = \text{rot } \Psi + \text{grad } \Phi$$

and the estimate is in $0(h^2)$ in H^{-1} (ω) norm, which is the natural norm for this quantity (see Theorem III.10). From a practical point of view, let us recall that the plate model solution is obtained as follows:

a) Φ is computed by a Crout type factorization of the matrix associated with problem (III.50).

b) the pair (Θ, Ψ) is then calculated for instance by using the elimination process mentioned hereafter. With self-evident notations, model (III.54) can be written in the following matrix form.

(IV.1)
$$\begin{cases} A\Theta + \overline{B}\Psi = G\Phi \\ B\Theta = 0 \end{cases}$$

where $\overline{B}$ is the transpose matrix of B. The following approximated system is associated with (IV.1)

(IV.2)
$$\begin{cases} A\Theta^\eta + \overline{B}\Psi^\eta = G\Phi \\ B\Theta^\eta = \eta M\Psi^\eta \end{cases}$$

where M is a weighting matrix (stands for a scalar product in the space $L^2(\omega)$) easy to invert. In practice, we choose the condensed mass, which leads to the system (after elimination of Ψ^η):

(IV.3)
$$\left(A + \frac{1}{\eta}\overline{B} M^{-1} B \right) \Theta^\eta = G \Phi \quad .$$

As mentioned in chapter II, the actual choice of parameter η is guided by Mindlin theory. By comparison with Mindlin model we are led to set:

$$\frac{1}{\eta} = \frac{E}{2(1+v)\,\varepsilon^2}$$

where (E, v) are respectively Young modulus and Poisson coefficient of the material used for the plate and ε half the thickness. System (IV.3) is numerically solved using the conjugate gradient algorithm preconditioned by the diagonal of matrix $\left[A + \frac{1}{\eta}\overline{B}\,M^{-1}\,B \right]$. A much more efficient method from the standpoint of C.P.U. time and fully using vector and parallel algorithms is described in the next sections IV.2 and IV.3. This will be the one we use for shells in Chapter X.

c) The deflection u_3 is computed by the downward-upward method since the matrix of the discretized system associated with $u_3{}^h$ and Φ^h is the same and was factorized during the first step of the solution.

IV.1.2   Numerical tests

Below we describe a comparison (in the sense of the solutions) between the mixed finite element scheme proposed herein and the QUAD4 suggested by Mac Neal. We also study the convergence of the mixed triangular element in situations where the QUAD4 is not appropriate.

IV.1.2.1 A square plate with simply supported edges, subjected to a uniform pressure

First of all, let us emphasize that the exact solution is known. The mechanical characteristics of the plate are reported on Figure IV.1. The results obtained with the so-called D.N.T. and D.N.Q. elements (Natural Duality for Triangles and Quadrangles, called "Dualité Naturelle" in French) and with the QUAD4 are shown on Figures IV.2, IV.3 and IV.4. The logarithm of the error with respect to the mesh size, denoted by h, is indicated. For each component the error is computed in $L^2(\omega)$ norm. A lot of integration points have been used in order to avoid a bad approximation of the analytic solution (seven points for triangles and nine for quadrangles). Convergence of the QUAD4 element is somewhat better concerning deflections and rotations. However, with the QUAD4, the accuracy on the bending moments is not as good and transverse shear is quite unsatisfactory. By contrast, the results obtained with the mixed element are relatively good for the bending moments and transverse shear. The local error was also evaluated in various points, which are the most often mentioned in the literature. The values obtained are represented in Figure IV.5. The rate of convergence, in $L^2(\omega)$ norm, with respect to the mesh was analyzed for the two families of elements.

Figure IV.1 - Mechanical characteristics of the plate

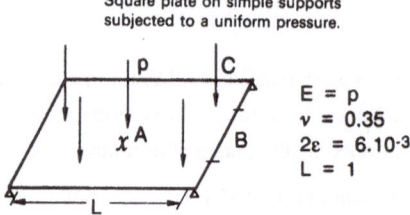

Figure IV.2 - Convergence results obtained with QUAD4

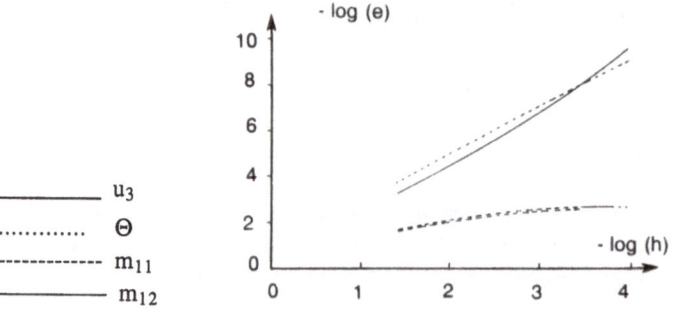

Figure IV.3 - Convergence results obtained with the D.N.T.

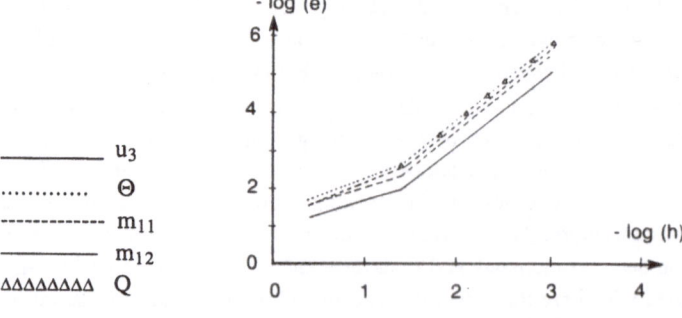

Figure IV.4 - Convergence results obtained with the D.N.Q.

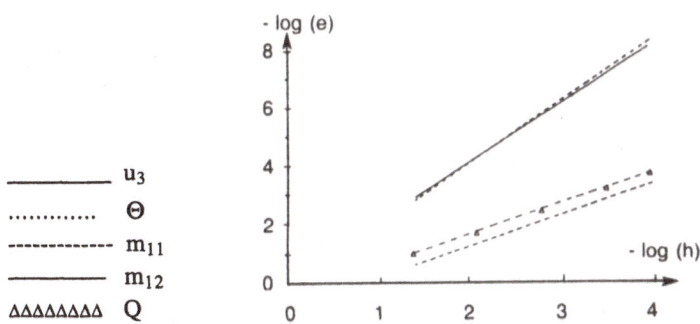

Figure IV.5 - Local error in points A, B and C (see Figure IV.1);
column 1 : QUAD4, (484 elements), column 2 : D.N.Q., (484 elements)

Quantites	u_3 (A)		θ_1 (B)		m_{11} (A)		m_{12} (C)		Q_1 (B)	
Elements	1	2	1	2	1	2	1	2	1	2
Relative error	5e-4	2e-4	2e-4	2.6e-3	7.3e-2	6e-4	1.3e-2	6e-3	?	5e-4

The conclusion is that convergence is slower for bending moments with QUAD4 whereas it is very good for deflection and rotations (see Figures IV.6 and IV.7).

Figure IV.6 - Comparison between D.N.Q. and QUAD4. Bending moment m_{11}

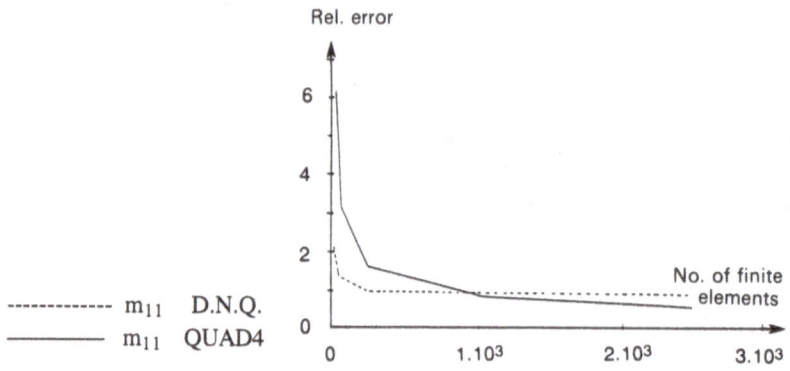

Figure IV.7 - Comparison between D.N.Q. and QUAD4. Bending moment m_{12}

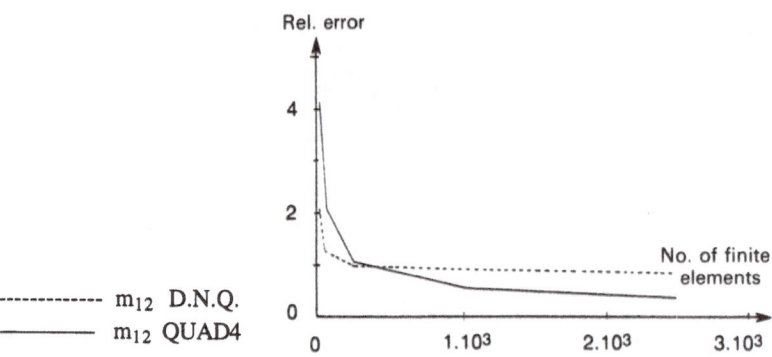

IV.1.2.2 Cantilevered plate subjected to a uniform pressure (the exact solution is unknown)

We consider a rectangular plate (see Figure IV.8). In this case QUAD4 converges poorly near the free edges. This agrees with the theoretical remarks of Chapter II where we detected a boundary layer phenomenon near the free edges. Here again, the results of the QUAD4 are not very good concerning the bending moments compared to the D.N.Q. (mixed elements). Transverse shear is represented on Figure IV.9. Only the mixed element gives an acceptable approximation. The comparisons are given in Figure IV.10 and IV.11 (the "exact" solution used as a reference for the comparison is obtained for a highly refined mesh).

Figure IV.8 - Mechanical characteristics of the plate

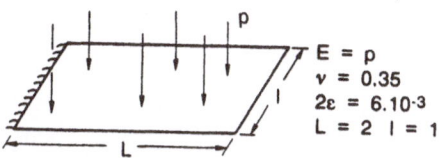

$E = \rho$
$\nu = 0.35$
$2\varepsilon = 6.10^{-3}$
$L = 2 \quad I = 1$

Figure IV.9 - Constant shear value obtained with D.N.Q.

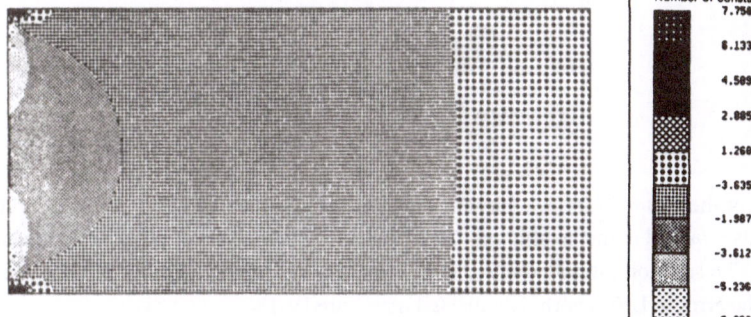
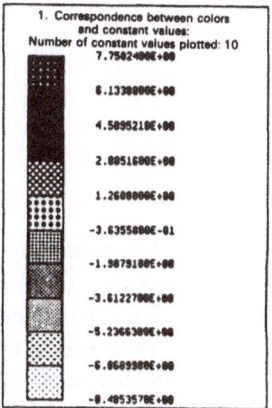

IV.1.2.3 Clamped plate with a hole in the center subjected to a uniform pressure (the exact solution is unknown)

Figure IV.10 - Convergence results obtained with QUAD4.

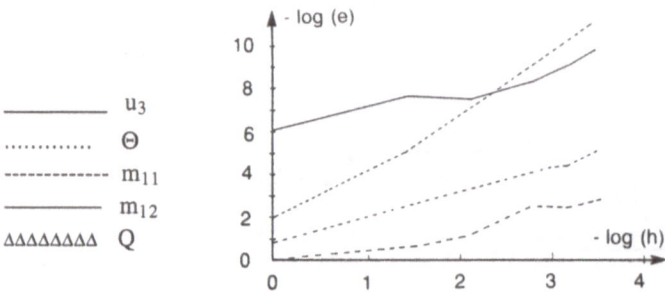

Figure IV.11 - Convergence results obtained with D.N.Q.

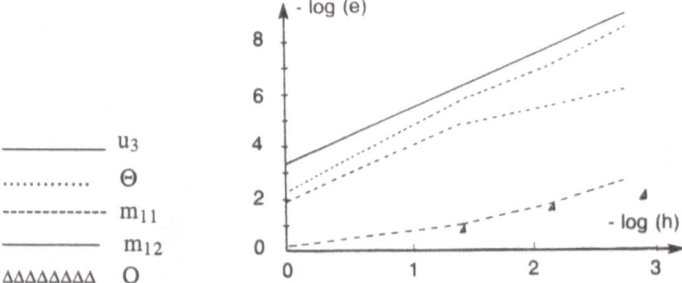

Considering the symmetries, we analyze a quarter of the plate (see Figure IV.12). The mesh refinements required in the neighbourhood of the curved edges are an incitation to use triangles (D.N.T.). The solution (assumed to be exact) has been calculated with a very fine mesh. The decay of the error in $L^2(\omega)$ norm is indicated on Figure IV.13.

The oscillations are due to the fact that the meshes are not included in one another (here again the logarithm of the error is shown with respect to the logarithm of the mesh size).

IV.1.2.4 Square plate with simply supported edges submitted to a pointwise load in the center (the exact solution is known)

The three elements (D.N.T., D.N.Q. and QUAD4) were used in this case of loading. The convergences are given on Figures IV.14, IV.15 and IV.16. The deflections and rotations are correctly approximated by all the elements, but QUAD4 raises a few small problems relative to the bending moments: the comparisons illustrating this are given on Figures IV.17, IV.18, IV.19 and IV.20.

Figure IV.12 - Quarter mesh of a clamped plate with a free edge hole in the center

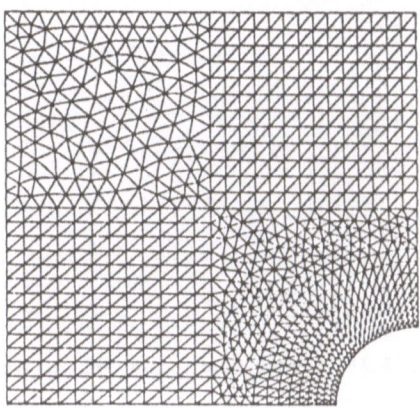

Figure IV.13 - Convergence results obtained with D.N.T.

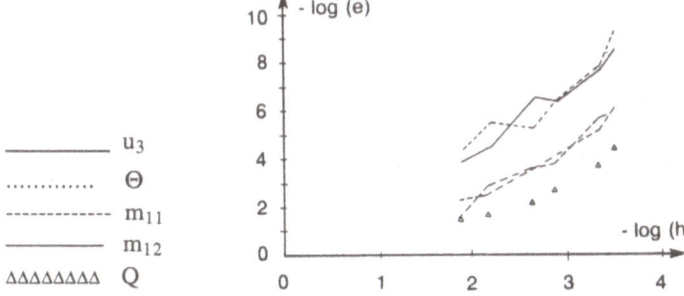

Figure IV.14 - Convergence results obtained with QUAD4

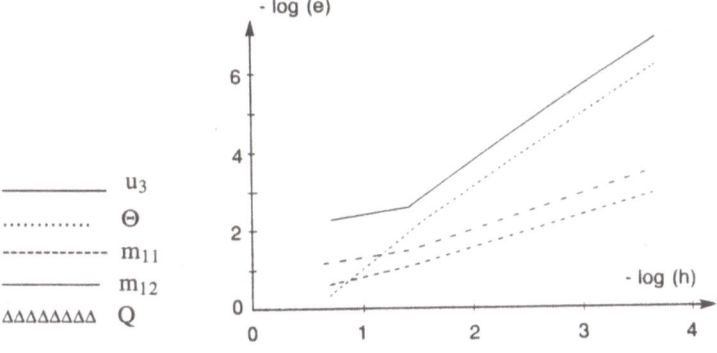

Figure IV.15 - Convergence results obtained with D.N.T.

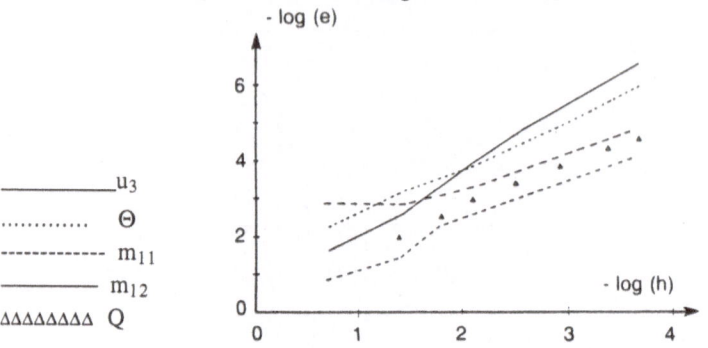

Figure IV.16 - Convergence results obtained with D.N.Q.

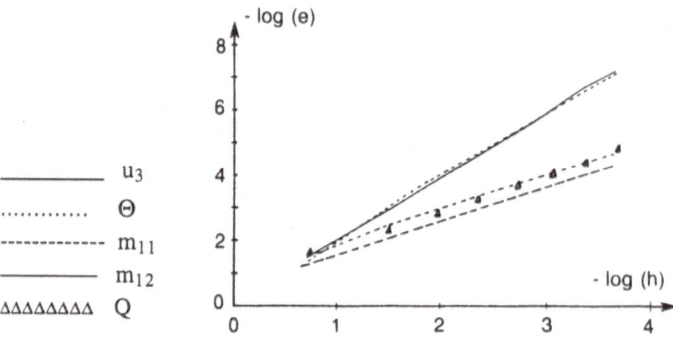

Figure IV.17 - Comparison between D.N.Q. and QUAD4

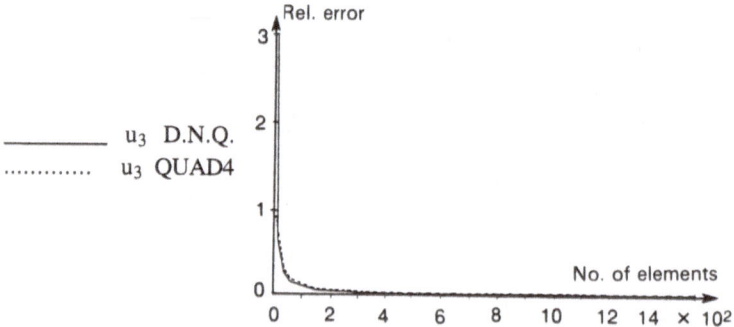

Figure IV.18 - Comparison between D.N.Q. and QUAD4

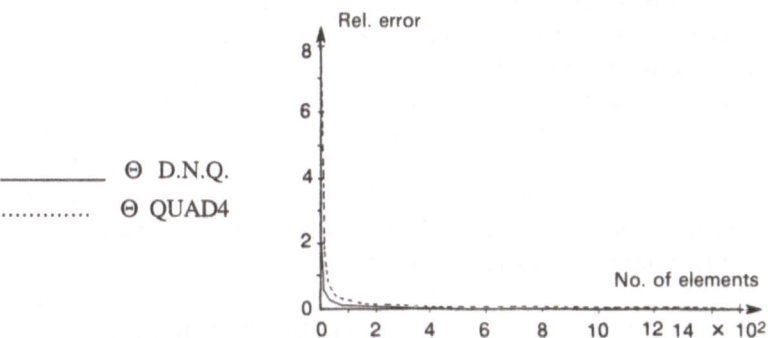

Figure IV.19 - Comparison between D.N.Q. and QUAD4

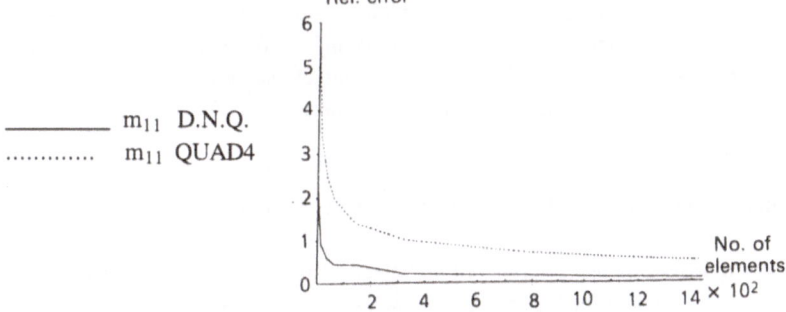

Figure IV.20 - Comparison between D.N.Q. and QUAD4

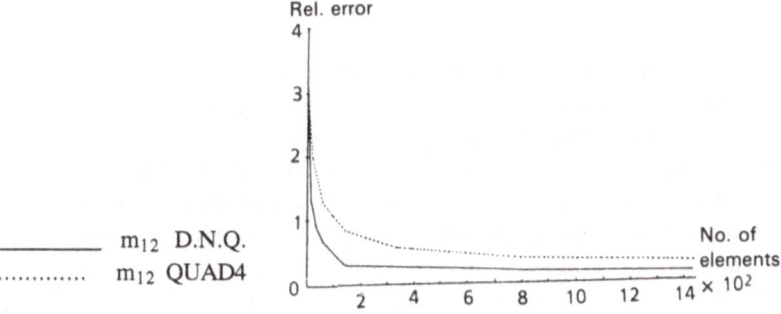

IV.1.3   *A few remarks relative to the numerical results*

First, it should be mentioned that we choose to express the error in $L^2(\omega)$ norm, rather than pointwise error, since its corresponds to the framework of theoretical analysis of the plate model. The abstract approaches to which we refer are detailed in Chapters II and III. In addition, comparisons other than those described herein were obtained in collaboration with J.L. Akian and Y. Ousset (in particular with D.K.T.). We refer the reader to this paper [2].

If we had to choose a universal element for plates, we would definitely refrain from recommending the family of mixed finite elements discussed herein. Everything depends on the criterion of choice and the use to be made of the element.

A few facts can however be mentioned. The results show that QUAD4 is efficient for deflection and rotations if the mesh is regular. In addition, it is very easy to use and is therefore a nice candidate for geometric nonlinearities. The D.N.T. and D.N.Q. elements are just as efficient but give better results for stresses. Such elements are therefore excellent candidates for non-linear analysis (geometry or materials). However, D.N.T. and D.N.Q. elements are a concept much more difficult to handle. Their use in computations requires a much higher level of know-how than QUAD4 (or any other element based on a primal formulation). Finally, in the analysis of edge effects or transverse shear, mixed elements are indubitably superior, which is confirmed by several analyses (the one due to J.L. Akian [1] is very interesting).

IV.2   Vectorial and parallel algorithms for mixed elements

The work described below applies to solving of the mixed system generated by the D.N.T. and D.N.Q. finite elements just described. But general conclusions can be drawn, extending to most mixed types of finite element systems. In effect, the effort was focused on solving the matrix system:

$$(IV.18) \qquad\qquad \begin{cases} A\Theta + \overline{B}\Psi = F \\ B\Theta = 0 \end{cases}$$

where A is a symmetric square matrix with dimensions (N_Θ, N_Θ) and matrix B is rectangular with dimensions (N_Ψ, N_Θ). In addition, it is recalled that $\overline{B}$ is the transpose matrix of B. Furthermore, N_Θ is very large compared with N_Ψ: in the applications, we have $N_\Theta \sim 22000$ and $N_\Psi \sim 3000$. All the numerical tests were conducted on an Alliant FX8/8 computer using multitasking but always remaining in the main memory. The available cache memory has a capacity of 512 Kbytes and the main memory has a capacity of 48 Mbytes. Each processor (called CE = Computer Element) has eight 32 words x 64 bytes vector registers. The algorithms below are written in Fortran 77 and make use of the parallel processing facilities allowed by the

Alliant compiler. It should be noted that similar results were obtained on a Stardent but with only two type P2 processors (as fast as six Alliant processors for this example).

IV.2.1   Three strategies for solving the system (IV.18)

IV.2.1.1   The dual variable elimination

This was the method we used in section IV. It consists in solving the system:

$$K^\eta \Theta^\eta = A + \frac{1}{\eta} \overline{B} . M^{-1} B \, \Theta^\eta = F$$

For this purpose, we used the conjugate gradient preconditioned by the diagonal. We evaluated Ψ^η by:

$$\Psi^\eta = \frac{1}{\eta} M^{-1} B \, \Theta^\eta$$

The main advantage of this strategy is the low memory capacity required : only Morse storage is used, i.e. only the non-zero terms are stored. In addition, matrix K^η is never assembled since the solving algorithm only requires matrix/vector products. This operation is then optimized for the architecture of the Alliant F8/8 computer. But it is necessary to use a double pointer system for Morse storage and the Alliant processors were not very performing for this type of manipulations of integers, contrary to the Convex computer which as a special processor for integers and the Stardent computer as well.

IV.2.1.2   The duality method

The principle of the duality algorithm is to eliminate (artificially!) the Θ variable in order to solve the problem in Ψ. This problem, called the dual problem, is written:

(IV.19)   minimize $\frac{1}{2} \left(B \, A^{-1} B \, \Psi, \Psi \right) - \left(B \, A^{-1} F, \Psi \right)$

   $\Psi \in R^{N_\Psi}$

The algorithm solution can be conducted with a conjugate gradient method requiring fewer iterations than the algorithm described in IV.2.1.1 (since $N_\Psi \ll N_\Theta$). There remains the problem of computing A^{-1} (which is not done). Two strategies were tested.

IV.2.1.2.1 Double gradient method

For each iteration of the gradient in Ψ, in solving (IV.19), A is inverted by another conjugate gradient method. As A is correctly conditioned (better than K^{η}!), the algorithm converges rapidly : we used diagonal, S.S.OR. and incomplete Cholesky factorization preconditioning for this (see [19]). However, there is still the problem of the double pointer system. Moreover, at each iteration, all the resolution has to be done.

IV.2.1.2.2 Factorization of A

We used Crout factorization of matrix A with the form :

$$A = L . D . \overline{L} .$$

D is the diagonal matrix and L is the lower triangular matrix with 1 on the diagonal. Each time the conjugate gradient is iterated in Ψ, it is then sufficient to carry out an upward-downward process with matrix L.

Much more memory is required by this method. But a single pointer is sufficient (skyline storage with bandwidth optimization). Two algorithms for reducing the surface of A are discussed below.

IV.2.1.2.3 Preconditioning of the dual problem

Three preconditionings were compared:

(i) $$\Delta_1 = \text{diag} \left\{ B \ \text{diag}^{-1}(A) \ \overline{B} \right\}^{\frac{1}{2}} ,$$

(ii) $$\Delta_2 = \text{diag} \left\{ B \ D^{-1} \ \overline{B} \right\}^{\frac{1}{2}}$$

where D is the diagonal matrix which appears in Crout factorization;

(iii) $$\Delta_3 = \text{diag} \left\{ B \ A^{-1} \ \overline{B} \right\}^{\frac{1}{2}} .$$

Four tests were conducted to evaluate the efficiency of these choices. They correspond to different geometries and boundary conditions. Table III gives the results obtained by the method where A is factorized. The first preconditioning appears to be the best from the standpoint of C.P.U. time, whereas the third generally gives the best results concerning the number of iterations; however, the C.P.U. time required makes it inoperative in practice.

It should be noted that the first preconditioning was also used for the double gradient method with identical results (whereas Δ_2 cannot be used unless A is factorized and Δ_3 is so long to compute that we only tested it when A was factorized). In Table III, the tests concerned:

(1) a square plate on simple supports subjected to a uniform pressure;
(2) a square plate subjected to a uniform pressure whose simply supported edges are respectively free, simply supported, clamped, simply supported;
(3) a cantilever plate subjected to a uniform pressure;
(4) a plate with a hole on simple support subjected to a uniform pressure.

Table III - Efficiency of the different preconditionings
on the conjugate gradient algorithm in Ψ

	N_Ψ		Precond. (none)	Precond. No. 1	Precond. No. 2	Precond. No. 3
Test No. 1	1,089	Iter	4	3	3	3
		tcpu	3.3	2.7	2.7	155.
Test No. 2	1,089	Iter	39	10	16	16
		tcpu	25.	7.	11.	166.
Test No. 3	2,178	Iter	65	32	116	32
		tcpu	85.	43.	149.	638.
Test No. 4	1,348	Iter	25	11	13	9
		tcpu	24.	11.	13.	289.

IV.2.2   Optimization of Crout factorization

Recall that the factorization algorithm is written ($A = L \cdot D \cdot \overline{L}$.):

$$D_j = A_{jj} - \sum_{k=1,j-1} L^2_{jk} D_k \qquad 1 \le j \le N_\Theta$$

$$L_{ij} = \left(A_{ij} - \sum_{k=1,j-1} L_{ik} D_k L_{jk} \right) / D_j \qquad 1 \le j \le i \le N_\Theta$$

For each subscript $j \le N_\Theta$, the computation of L_{ij}, where $j + 1 \le i \le N_\Theta$, requires knowing D_k, $1 \le k \le j$, and L_{pq} where : $j \le p \le N_\Theta$, $1 \le q \le j - 1$.

This means that once D_j has been computed, the L_{ij} terms can be computed (with j fixed) using parallelism on subscript i from $j + 1$ to N_Θ. Matrix A is thus factorized column by column, in parallel. In addition, expressions like :

$$\sum_{k = 1, j - 1} L_{ik} D_k L_{jk}$$

are vectorized by the Fortran compiler.

To optimize the use of the cache memory, local subtables were introduced to store the frequently used auxiliary data. For instance, the diagonal of A is stored in a particular vector during factorization and is not introduced in place of the diagonal of L until the outcome of this operation. An important point is that all the tests are made on 64 bits.

The performance of the factorization algorithm is described below. First, the influence of parallelism is examined (see Figure IV.21). The number of processors varies from one to eight. In the last case, the acceleration is by a factor of 6.7, compared to one processor, whereas the size of A is 1 Megabyte.

Figure IV.21 - Influence of the number of processors on factorization

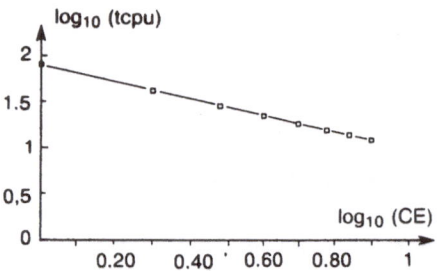

During the second phase, we examined the C.P.U. time according to the surface of matrix A to be factorized (see Figure IV.22). The performance is maintained satisfactorily; in particular, saturation of the cache memory (512 Kbytes) does not significantly affect the results.

Figure IV.22 - Variation of factorization time versus the surface of the matrix A

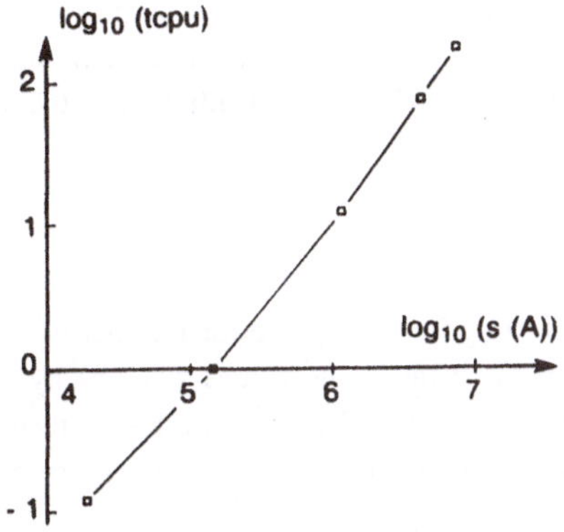

IV.2.3   Optimization of node renumbering

IV.2.3.1 Gibbs-Poole-Stockmeyer algorithm (see reference [19])

This algorithm is a variant of the Cuthill-Mac Kee algorithm whose two main ideas are as follows:

"when a node is numbered, the adjacent nodes are numbered by increasing degrees (the degree of a node is the number of its neighbors); insofar as possible, the first node must have the minimum degree".

A few definitions are recalled below to describe the Gibbs-Poole-Stockmeyer variant in greater detail.

Definition 1: *Levels of a graph initialized in node "v"*
These are sets of nodes denoted L_i defined by:
$L^1 = \{v\}$ (singleton);

$$L_i = \left\{ \text{set of nodes adjacent to } L_{i-1} \text{ which are not counted in } \bigcup_{k=1, i-1} L_k \right\}$$

Definition 2: *Depth of a graph initialized in node "v"*
The depth of the graph initialized in node "v" is the number of graph levels.

Definition 3: *Width of a graph*

The width of a graph is the maximum of the cardinals of sets L_i, $i \geq 1$.

The first step of the G.P.S. algorithm consists in finding a graph with a minimum width among the graphs with a maximum depth. The second consists in numbering the nodes level by level using the same principles as in Cuthill – MacKee algorithm. Note that the algorithm is run only once.

IV.2.3.2   Algorithm of the mean

This is an iterative algorithm. Starting from a mesh node numbering, an indicator is associated with each node. This indicator is the mean of the values taken on by an increasing function of the numbers of its neighbors, including itself. The mesh nodes are then numbered by increasing order of the indicators. Obviously, the starting point and the choice of the increasing function play an important role.

In the test conducted, this algorithm can give better results than the one described in IV.2.3.1., except when the mesh mixes triangles and quadrangles but even with a very simple (affine!) increasing function, the algorithm requires several tens of iterations, which is very costly in C.P.U. time. For this reason, the G.P.S. algorithm was chosen and optimized from the vectorial and parallel standpoint. The Figure IV.23 shows the C.P.U. times of the G.P.S. algorithm according to the number of processors used (3201 nodes are renumbered in this example).

Figure IV.23 - G.P.S. Algorithm computation time versus number of processors

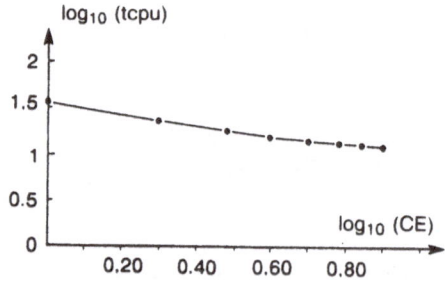

Table IV - Comparative performance of the three strategies

CPU time in seconds	Method 1	Method 2	Method 3
Pointers	260.	260.	41.3
Renumbering	–	–	11.1
Factorization of A	–	–	14.6
Computation of ψ	0.	544.	11.1
Computation of θ	1,956.	45.	0.9
TOTAL	2,216.	849.	79.0

Table V - Detail of C.P.U. time of the third strategy

CPU time in seconds	Test No. 1	Test No. 2	Test No. 3
N_θ	12,870	15,402	6,402
$N\psi$	2,178	2,601	1,089
$S(A)$	2,220,517	4,299,939	1,135,431
Pointers	12.6	17.0	4.0
Renumbering	20.6	45.3	12.5
Factorization	21.7	71.1	12.3
ψ	42.5	45.5	7.1
θ	1.2	2.5	0.6

Since the algorithm is strongly recursive, the acceleration factor is only 2.9 for eight processors. Actually, it will be seen below (see Table V) that renumbering is one of the largest steps in solving the plate model proposed.

IV.2.4   *Numerical tests*

Table IV gives the comparisons of three computation strategies (see sections IV.2.1) for solving a plate model problem : the case of a plate with a hole on a simple support subjected to a uniform pressure . Let us recall that the mesh consists of quadrangles and triangles (see Figure IV.24).

Figure IV.24 - Mesh of the test plate

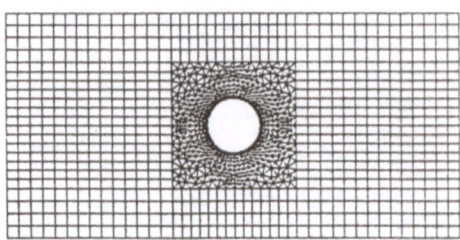

The dimensions of the problem are $N_\Theta = 8968$ and $N_\psi = 1348$, surface (A) = 123 Kbytes in Morse storage and surface (A) = 1.37 Megabytes in skyline storage. The three strategies are recalled for reference.

Strategy 1 – Solution of

$$\left[A + \frac{1}{\eta} \overline{B} M^{-1} B \right] \Theta^\eta = F$$

using the conjugate gradient algorithm preconditioned by the diagonal. There is no assembly of the complete matrix and A is stored in Morse form.

Strategy 2 – A duality algorithm is used. For each iteration, the primal problem is solved by a conjugate gradient algorithm. The preconditioning is that defined by Δ_1 (see section IV.2.1.2.3) for the dual problem and by the diagonal of A for the primal problem. Matrix A is again stored in Morse form.

Strategy 3 – A duality algorithm is used; each iteration, the primal problem is solved by a direct method (Crout factorization of A). The preconditioning of the duality algorithm is that associated with Δ_1 and the surface of A is minimized by the G.P.S. algorithm described in section IV.2.3.1.

The results are summarized in Table IV and show the indubitable superiority of the third strategy concerning C.P.U. time.

Table V gives the detail of the main steps (in C.P.U. time) of the third strategy for the following tests:

1) Cantilever plate subjected to a uniform pressure;

2) Square plate subjected to a uniform pressure, whose edges are respectively free, simply supported, clamped, free.

3) Square plate on simple support subjected to a uniform pressure.

IV.3   Concluding remarks

The numerical aspects of the mixed finite element method described herein can be summarized as follows.

Transverse shear, which is the Lagrangian multiplier of the Kirchhoff-Love kinematical equation, is broken down into the sum of two potentials denoted Φ and Ψ in the text. Φ is computed by a Laplacian decoupled from the other unknowns. In addition, potential Ψ associated with rotation Θ of the normal to the plate are solutions of a Stokes model (Ψ acts as the pressure and Θ as the velocity up to a rotation of $\frac{\pi}{2}$). Finally the deflection u_3 is obtained by solving the problem characterizing Φ except for the right-hand side.

The numerical performances as concerns both accuracy and computation time show that this type of element is superior to Mac Neal QUAD4 element.

The initial incentives which led to developing these numerical methods were multi-layered composite materials for which transverse shear plays an important role. The formulation we have given herein is of course situated in this framework. However, the numerical results are limited to the isotropic case in order to facilitate comparisons with the literature on thin plate elements subjected to bending. Further numerical results are given in the next chapter which is devoted to multi-layered composite plates.

REFERENCES

[1] AKIAN J.L., [1988], Méthodes d'éléments finis mixtes pour le calcul des contraintes dans un composite au niveau des trous chargés, Rapport technique No 37/4242 RY 062 R, ONERA.

[2] AKIAN J.L., DESTUYNDER Ph., NEVERS Th., OUSSET Y., [1988],Quelques méthodes d'éléments finis mixtes pour les plaques minces, CNRS-GRECO Calcul des Structures (Ed. Pluralis), Vol. 2.

[3] BELYTSCHKO T., LASRY D., [1987], Transverse shear oscillations in four node quadrilateral elements, Comput. & Structures, 27, p. 393-398.

[4] BELYTSCHKO T., STOLARSKI M., [1983], Shear and membrane locking in curved C^0 elements, Comput. Methods Appl. Mech. Engrg., 41, p. 279-288.

[5] BERCOVIER M., [1978], Perturbation of mixed variational problems. Applications to mixed finite elements methods, RAIRO, R2, 12, p. 211-228.

[6] DESTUYNDER Ph., NEVERS Th., [1988], A new finite element scheme for bending plates, Comput. Methods Appl. Mech. Engrg., 68, p. 127-139.

[7] DESTUYNDER Ph., NEVERS Th., [1990], Some numerical aspects of mixed finite elements for bending plates, Comput. Methods Appl. Mech. Engrg., 78, p. 73–87.

[8] GALLIVAN K., JALBY W., MEIER U., [1987], The use of BLAS 3 in linear algebra on a parallel processor with a hierarchical memory, SIAM J. Sci. Stat. Comp.

[9] GIBBS N.E., POOLE W.G., STOCKMEYER P.K., [1976], An algorithm for reducing the bandwidth and profile of a sparse matrix, SIAM J. Numer. Ana., Vol. 13, No. 2.

[10] HUGHES T.J.R., TAYLOR R.L., KANOKNUKULCHOI W., [1977], A simple and efficient element for plate bending, Int. J. Numer. Methods Engrg, Vol. 2, 10, p. 1529-1543.

[11] HUGHES T.J.R., TEZDUYAR T.E., [1981], Finite elements based upon Mindlin plate theory with particular reference to the four node bilinear isoparametric element, ASME J. Appl. Mech., 46, p. 587-597.

[12] MAC NEAL R.H., [1978], A simple quadrilateral shell element, Comput. & Structures, 8, p. 175-183.

[13] MINDLIN R.D., [1951], Influence of rotary inertia and shear on flexural motion of

isotropic elastic plates, J. Appl. Mech., 18, p. 31-38.

[14] NEVERS Th., [1986], Modélisation théorique et numérique du délaminage des plaques composites, Doctoral Dissertation, École Centrale Paris.

[15] NEVERS Th., SALAÜN M., [1989], Optimization of mixed formulations on a parallel computer, Proceedings of the 5th International Symposium on numerical methods in engineering. Computational Mechanics Publications, p. 123-129, Springer-Verlag, Berlin.

[16] PARISCH H., [1981], Large displacements of shells including material nonlinearities, Comput. Methods Appl. Mech. Engrg., 27, p. 183-214.

[17] ROUX F.X., [1989], Accélération de la convergence par reconjugaison des directions de descente d'une méthode de résolution par sous-domaines d'un problème d'élasticité linéaire, CRAS, Paris, 308, série 1, p. 193-198.

[18] TAYLOR C., HOOD P., [1974], Navier-Stokes equations using mixed interpolation, in Finite element in flow problem, Oden Ed. U.A.H. Press.

[19] LASCAUX P., THEODOR R., [1993], Analyse numérique matricielle appliquée à l'art de l'ingénieur, Masson, Paris.

Chapter 5

A NUMERICAL MODEL FOR DELAMINATION OF COMPOSITE MULTILAYERED PLATES

V.0   A brief description of the chapter

After a brief recall on the laminated plate theory, a computational model for studying the delamination is presented. In order to avoid a too complicated mathematical justification (which is included in [1]), we base our developments on physical feelings. Then a mechanical example including comparison with experiments is discussed. All the chapter is a survey of a study performed five years ago with our friend Thierry Nevers. The extended versions can be found in the references [2], [3], [4], [5], [6].

V.1   What is delamination of thin multilayered plates

With the development of multi-layered composite materials, a new damage mechanism appeared in thin plates and shells. It already limits the range of applicability of such structures unless a huge step can be done in the understanding and then the prediction of it. This is the delamination. Usually it appears to be initiated by an overstressing near the edges (even, and maybe – in a way – mainly the free edges). Then a planar crack separating two layers of the laminate develops inward the plate. Its evolution involves several micro-mechanisms like fracture of fibers, decohesion between the adhesive and the fibers, voids growth in the adhesive. But, as far as the delaminated area is large enough, one can try to apply fracture mechanics based on energy considerations. More precisely, if we assume that the crack evolution only depends on the delamination curve, the energy release rate is defined as the derivative of the elastic stored energy with respect to the position of this curve. Thus it appears as a thermodynamic force, and one can try to use the linear fracture mechanics based on a yield criterion (following the idea of A. Griffith [7]). This is the goal of this chapter, which is an application of the theory developed in the preceding ones.

The first section V.2 is devoted to a mechanical introduction where the main fracture mechanics concepts are presented. The purpose is not to discuss a basic theory on the delamination of composite multilayered structures, but only to precise a few notations. These concepts are the framework that we are dealing with in this Chapter. For instance, the birth of the damage mechanism is not discussed. Let us mention in this direction the works of S.S. Wang [8], A.S.D. Wang [9], F. Crossman [10], I.S. Raju - J.H. Crews [11] and L. Anquez – A. Bern –

J. Renard [12]. The mathematical aspects of overstressing developed near the free edges and which is involved in the birth of the delamination has been discussed in the context of a boundary layer theory by J.L. Davet – Ph. Destuynder [13], Y. Ousset [14] and Ph. Destuynder – Y. Ousset [15]. A parameter study of the stress singularities in the edge effect has been done in Ph. Destuynder – E. Bonnet – E. Michelin – B. Templier [16]. With another respect, the method that we describe here has been extended to more general applications by Y. Ousset [17] (coupling with homogeneization) and by Y. Ousset – F. Roudolph [18] (taking into account several delamination areas and analyzing their effects).

Figure V.1 - A delaminated composite plate

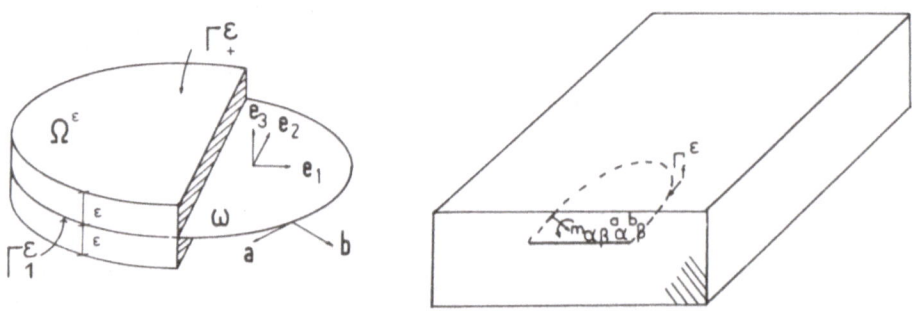

V.2   The three-dimensional multilayered composite plate model with delamination

Let us consider a thin plate as shown on figure V.1. The thickness is 2 ε and the medium surface is denoted by ω. The whole plate is divided into three sub-parts (see Figure V.1). One is Ω^ε_s and corresponds to the "safe" portion of the structure (i.e. without delamination). The other two – say Ω^ε_u and Ω^ε_l – are respectively the **upper** part and the **lower** part of the delaminated portion. The separation surface – say Γ^ε_f – is assumed to be planar. Both displacements and normal stresses are continuous across Γ^ε_f. Therefore the Principle of Virtual Work (see Chapter I) is formulated as follows:

$$(V.1) \qquad \forall\ v \in V^\varepsilon,\ \int_{\Omega^\varepsilon} \sigma_{ij}\gamma_{ij}(v) = \int_{\Gamma^\varepsilon_l} h_i v_i\ +\ \int_{\Gamma^\varepsilon_\pm} g^\pm_i v_i$$

where $\Omega^\varepsilon = \Omega^\varepsilon_s \cup \Omega^\varepsilon_u \cup \Omega^\varepsilon_l$, h_i being the surface forces applied on the lateral part of

Ω^ε denoted Γ_1^ε. The transverse surface forces applied on the upper and lower surfaces of the plate are denoted by $g_i^\pm$ as in the previous chapter. The new feature comes from the constitutive relationship between strains and stresses. The material is assumed to be monoclinic (which is perfectly correct for a multilayered plate made of differently oriented fibers depending on the layer). Therefore the relationship is:

(V.2) $$\sigma_{ij} = R_{ijkl} \gamma_{kl}(u)$$

where the stiffness tensor satisfies:

$$\begin{cases} R_{ijkl} = R_{jikl} \quad \forall \; i,j,k,l \in \{1,2,3\} \\ R_{ijkl} = R_{klij} \qquad " \qquad\qquad " \\ R_{\alpha 333} = R_{\alpha\beta\gamma 3} = 0 \; \forall \; \alpha,\beta,\gamma \in \{1,2\} \end{cases}$$

The last condition traduces the property of the material to be monoclinic. Hence, the equation (V.2) can be split into the following relations:

(V.3) $$\begin{cases} \sigma_{\alpha\beta} = R_{\alpha\beta\mu\lambda} \, \gamma_{\mu\lambda}(u) + R_{\alpha\beta 33} \, \gamma_{33}(u), \\ \sigma_{\alpha 3} = 2 R_{\alpha 3\beta 3} \, \gamma_{\beta 3}(u), \\ \sigma_{33} = R_{3333} \, \gamma_{33}(u) + R_{33\alpha\beta} \, \gamma_{\alpha\beta}(u). \end{cases}$$

It is then possible to invert these formulae in order to express the strains $\gamma_{\alpha\beta}(u)$ with respect to the stress $\sigma_{\alpha\beta}$. A simple calculation leads to the following equations, involving the so-called compliance tensor S_{ijkl} which is the inverse of R_{ijkl}:

$$\gamma_{ij}(u) = S_{ijkl} \, \sigma_{kl}$$

or else:

$$\begin{cases} \gamma_{\alpha\beta}(u) = S_{\alpha\beta\mu\lambda} \, \sigma_{\mu\lambda} + S_{\alpha\beta 33} \, \sigma_{33}, \\ \gamma_{\alpha 3}(u) = 2 S_{\alpha 3\beta 3} \, \sigma_{\beta 3}, \\ \gamma_{33}(u) = S_{3333} \, \sigma_{33} + S_{33\alpha\beta} \, \sigma_{\alpha\beta}. \end{cases}$$

It is worth pointing out that, unless $S_{\alpha\beta 33}$ would be zero, the term $S_{\alpha\beta\mu\lambda}$ is not the inverse tensor of $R_{\alpha\beta\mu\lambda}$. Hence, the inverse of $S_{\alpha\beta\mu\lambda}$, which is therefore different from $R_{\alpha\beta\mu\lambda}$, is denoted, in the rest of the chapter, by $\overline{R}_{\alpha\beta\mu\lambda}$.

Let us summarize the obtained three-dimensional model. It consists in finding a displacement field u such that:

$$
\text{(V.4)}
\begin{cases}
u \in V^{\varepsilon} = \left\{ v \mid v = (v_i) \in H^1(\Omega^{\varepsilon}); \; v_i = 0 \text{ on } \Gamma^{\varepsilon}_0 \right\} \\[2mm]
+ 4 \int_{\Omega^{\varepsilon}} R_{\alpha 3 \beta 3} \, \gamma_{\alpha 3}(u) \, \gamma_{\beta 3}(v) + \int_{\Omega^{\varepsilon}} R_{\alpha \beta 33} \, \gamma_{\alpha \beta}(u) \, \gamma_{33}(v) \\[2mm]
\quad + \gamma_{33}(u) \, \gamma_{\alpha \beta}(v) + \int_{\Omega^{\varepsilon}} R_{3333} \, \gamma_{33}(u) \, \gamma_{33}(v) \\[2mm]
= \int_{\Gamma^{\varepsilon}_1} h_i \, v_i + \int_{\Gamma^{\varepsilon}_{\pm}} g_i \, v_i \quad .
\end{cases}
$$

Obviously a more general loading case could be considered, for instance with in plane loading like traction on the lateral boundary of the plate. But it is not necessary, for the only understanding of the theoretical principles used here, to complicate the basic model under discussion. Furthermore, the existence and uniqueness of a solution, to the variational equation (V.4), is directly obtained from the general result of Chapter I.

V.3   A plate model for large delamination

The plate model is derived from the three-dimensional theory in the same manner as we did in Chapter I. More precisely, because of the delaminated area, the plate is divided into three sub-plates (say Ω^{ε}_s , Ω^{ε}_u and Ω^{ε}_l) as shown on Figure V.1. On each one, the kinematical Kirchhoff-Love assumption is assumed. Thus one has:

$$
\gamma_{i3}(u) = 0 \qquad \text{for } i = 1,2,3
$$

but this relation is only true inside the sub-plates and not at the intersection between them where a boundary layer appears.

A simple computation on each sub-plate from enables one to derive the following expression of the displacement field.

$$(V.5) \begin{cases} \textbf{on } \omega_s \text{ (see Figure V.2)} \\ \quad \boxed{\begin{array}{l} \bullet \ u_3 \text{ is } x_3 \text{ independent} \\ \bullet \ u_\alpha = \underline{u}_\alpha - x_3 \, \partial_\alpha u_3 \,, \ \alpha = 1, 2 \text{ where } \underline{u}_\alpha \text{ is } x_3 \text{ independent} \end{array}} \\ \textbf{on } \omega_d \text{ (see Figure V.2)} \\ \quad x_3 > z \\ \quad \boxed{\begin{array}{l} \bullet \ u^+_3 \text{ is } x_3 \text{ independent} \\ \bullet \ u^+_\alpha = \underline{u}^+_\alpha - \left(x_3 - \xi^+\right) \partial_\alpha u^+_3 \,, \text{ where } \underline{u}^+_\alpha \text{ is } x_3 \text{ independent and } \xi^+ = (\varepsilon + z)/2 \end{array}} \\ \quad x_3 < z \\ \quad \boxed{\begin{array}{l} \bullet \ u^-_3 \text{ is } x_3 \text{ independent} \\ \bullet \ u^-_\alpha = \underline{u}^-_\alpha - \left(x_3 - \xi^-\right) \partial_\alpha u^-_3 \,, \text{ where } \underline{u}^-_\alpha \text{ is } x_3 \text{ independent and } \xi^- = (z - \varepsilon)/2 \end{array}} \end{cases}$$

At the interface between the three sub-plates, the displacement continuity implies:

$$\boxed{\begin{array}{ll} \text{(i)} & u^+_3 + u^-_3 = u_3 \\ \text{(ii)} & \underline{u}^+_\alpha + \xi^+ \, \partial_\alpha u^+_3 = \underline{u}^-_\alpha + \xi^- \, \partial_\alpha u^-_3 = \underline{u}_\alpha \,, \\ \text{(iii)} & \partial_\alpha u^+_3 = \partial_\alpha u^-_3 = \partial_\alpha u_3 \end{array}}$$

or else

$$(V.6) \quad \boxed{\begin{array}{ll} u^+_3 = u^-_3 = u_3 & \partial_\alpha u^+_3 = \partial_\alpha u^-_3 = \partial_\alpha u_3 \\ \underline{u}^+_\alpha = \underline{u}_\alpha - \xi^+ \, \partial_\alpha u_3 & \underline{u}^-_\alpha = \underline{u}_\alpha - \xi^- \, \partial_\alpha u_3 \end{array}}$$

Figure V.2 - The geometry of the delaminated plate

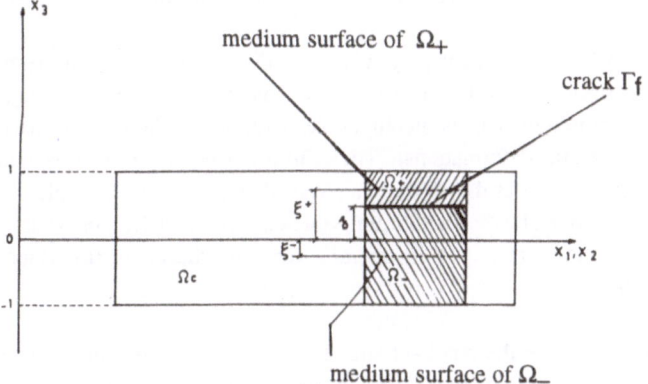

Remark V.1.

The above relations (V.6) imply that no opening mode is permitted at the crack tip. This is due to the fact that the two lips of the crack (upper and lower plates) have to remain tangential during deformation at the connecting point (see Figure V.3).

Figure V.3 - Deformation of the delaminated plate

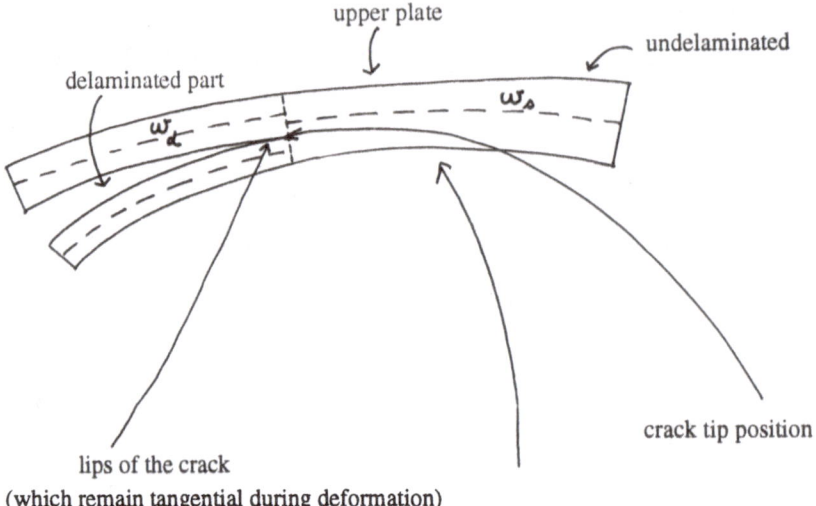

upper plate

undelaminated

delaminated part

crack tip position

lips of the crack
(which remain tangential during deformation)

But these considerations have to be carefully interpreted. As a matter of fact the normal stress (in the direction of the thickness of the plate) is given by (taking into account the kinematical hypothesis):

$$\sigma_{33} = R_{33\alpha\beta}\,\gamma_{\alpha\beta} = R_{33\alpha\beta}\,\partial_\alpha u_\beta - x_3\,R_{33\alpha\beta}\,\partial_{\alpha\beta}u_3$$

and thus $\sigma_{33} \neq 0$ even if $\gamma_{33}(u) = 0$. One could also object that the previous expression for σ_{33} is not really valid. If we refer to Chapter I, we mentioned that the three-dimensional constitutive relationship is incompatible with the Kirchhoff-Love assumptions (kinematical hypothesis on the deformations). The equilibrium equations led to another expression which is more reliable as far as they obey to a physical principle, the Principle of Virtual Work which is locally interpreted by the equilibrium equations. Nevertheless, the stress component σ_{33} is not zero and thus there exists an opening force on the crack tip which stimulates the crack propagation.   ■

Let us now define the resultant stresses and bending moments in each sub-plate. Thus we introduce the resulting stress components (membrane stresses), and the bending moments on each sub-plate by the following expressions:

on ω_s :

$$n_{\alpha\beta} = \int_{-\varepsilon}^{\varepsilon} \sigma_{\alpha\beta} , \qquad\qquad m_{\alpha\beta} = \int_{-\varepsilon}^{\varepsilon} x_3 \sigma_{\alpha\beta} ,$$

(V.7)   **on ω_d :**

$$
\begin{cases}
n^+{}_{\alpha\beta} = \int_{z}^{\varepsilon} \sigma_{\alpha\beta} , & m^+_{\alpha\beta} = \int_{z}^{\varepsilon} \left(x_3 - \xi^+\right)\sigma_{\alpha\beta} , \\[4mm]
n^-{}_{\alpha\beta} = \int_{-\varepsilon}^{z} \sigma_{\alpha\beta} , & m^-_{\alpha\beta} = \int_{-\varepsilon}^{z} \left(x_3 - \xi^-\right)\sigma_{\alpha\beta} .
\end{cases}
$$

Then the constitutive relationships are written on each subplate. If the plan stress stiffness tensor is denoted by: $\overline{R}_{\alpha\beta\mu\lambda}$ (it is the inverse of the in-plan compliance tensor denoted by $S_{\alpha\beta\mu\lambda}$), we introduce the terms:

$$
\begin{aligned}
R^M_{\alpha\beta\mu\lambda} &= \int_{-\varepsilon}^{\varepsilon} \overline{R}_{\alpha\beta\mu\lambda} , \quad R^{M+}_{\alpha\beta\mu\lambda} = \int_{z}^{\varepsilon} \overline{R}_{\alpha\beta\mu\lambda} , \quad \overline{R}^{M-}_{\alpha\beta\mu\lambda} = \int_{-\varepsilon}^{z} \overline{R}_{\alpha\beta\mu\lambda} , \\[4mm]
R^F_{\alpha\beta\mu\lambda} &= \int_{-\varepsilon}^{\varepsilon} x^2_3\, \overline{R}_{\alpha\beta\mu\lambda} , \quad R^{F+}_{\alpha\beta\mu\lambda} = \int_{z}^{\varepsilon} \left(x_3 - \xi^+\right)^2 \overline{R}_{\alpha\beta\mu\lambda} , \\[4mm]
\overline{R}^{F-}_{\alpha\beta\mu\lambda} &= \int_{-\varepsilon}^{z} \left(x_3 - \xi^-\right)^2 \overline{R}_{\alpha\beta\mu\lambda} , \\[4mm]
R^{C+}_{\alpha\beta\mu\lambda} &= \int_{z}^{\varepsilon} \left(x_3 - \xi^+\right) \overline{R}_{\alpha\beta\mu\lambda} , \quad \overline{R}^{C-}_{\alpha\beta\mu\lambda} = \int_{-\varepsilon}^{z} \left(x_3 - \xi^-\right) \overline{R}_{\alpha\beta\mu\lambda}
\end{aligned}
$$

(V.8)

where $\xi^+ = \dfrac{\varepsilon + z}{2}$ and $\xi^- = \dfrac{-\varepsilon + z}{2}$ are the coordinates along the axis x_3 of the medium surface of each sub-plate corresponding to the delaminated area.

Finally the strain-stress constitutive equations are explicited as follows:

(V.9)

$$\left\{ \begin{array}{l}
\text{on } \omega_s : \\[4pt]
\boxed{n_{\alpha\beta} = R^F_{\alpha\beta\mu\lambda}\gamma_{\mu\lambda}(\underline{u}), \quad m_{\alpha\beta} = -R^F_{\alpha\beta\mu\lambda}\partial_{\mu\lambda}u_3} \\[10pt]
\text{on } \omega_d : \\[4pt]
\boxed{\begin{array}{l}
n^{\pm}_{\alpha\beta} = R^{M\pm}_{\alpha\beta\mu\lambda}\gamma_{\mu\lambda}(\underline{u}^{\pm}) - R^{C\pm}_{\alpha\beta\mu\lambda}\partial_{\mu\lambda}u_3 \\[8pt]
m^{\pm}_{\alpha\beta} = -R^{F\pm}_{\alpha\beta\mu\lambda}\partial_{\mu\lambda}u_3 + R^{C\pm}_{\alpha\beta\mu\lambda}\gamma_{\mu\lambda}(\underline{u}^{\pm})
\end{array}}
\end{array} \right.$$

From the three-dimensional principle of Virtual Work and by limiting the virtual displacement to piecewise Kirchhoff-Love kinematical fields, one obtains (see Chapter I)

(V.10)

$$\left\{ \begin{array}{l}
\forall \, (v_\alpha), \displaystyle\int_{\omega_s} n_{\alpha\beta}\,\partial_\alpha v_\beta + \int_{\omega_d} n^+_{\alpha\beta}\,\partial_\alpha v^+_\beta + \int_{\omega_d} n^-_{\alpha\beta}\,\partial_\alpha v^-_\beta = F_\alpha(v_\alpha) \\[14pt]
\forall \, (v_3), -\displaystyle\int_{\omega} m_{\alpha\beta}\,\partial_{\alpha\beta} v_3 - \int_{\omega_d} m^+_{\alpha\beta}\,\partial_\alpha v_3 - \int_{\omega_d} m^-_{\alpha\beta}\,\partial_\alpha v_3 = F_3(v_3)
\end{array} \right.$$

where (v_α, v_3) are displacement fields satisfying the boundary conditions and only depending on the coordinates (x_1, x_2) varying in the medium surface of the plate. Let us refer to formulae (V.5). Furthermore the right-hand sides are respectively defined by the following formulae where we assume for sake of simplicity that no force is applied on the delaminated part of the plate:

$$\left\{ \begin{array}{l}
F_\alpha(v_\alpha) = \displaystyle\int_{\gamma_1} \left[\int_{-\varepsilon}^{\varepsilon} h_\alpha \right] v_\alpha + \int_\omega (g^+_\alpha + g^-_\alpha)\, v_\alpha , \\[20pt]
F_3(v_3) = \displaystyle\int_{\gamma_1} \left[\int_{-\varepsilon}^{\varepsilon} h_3 \right] v_3 - \int_{\gamma_1} \left[\int_{-\varepsilon}^{\varepsilon} x_z h_\alpha \right] \partial_\alpha v_3 , \\[20pt]
\quad + \displaystyle\int_\omega (g^+_3 + g^-_3)\, v_3 - \int_\omega [\varepsilon(g^+_\alpha - g^-_\alpha)]\, \partial_\alpha v_3 .
\end{array} \right.$$

The existence and uniqueness of a solution to (V.6), (V.7), (V.8), (V.9) and (V.10) can be obtained by the same method as the one we introduced in Chapter I. Let us recall that it is derived from the properties of the three-dimensional model and that it is not necessary to make explicit the plate model. The characterization of the displacement field subspace is the one given in (V.5). For additional information we refer the reader to section I.3.1. in Chapter I.

A basic point is the computation of the resulting transverse shear stress which plays a very important role in the delamination process.

Following the ideas introduced in chapter I, we use the local equilibrium equations in order to characterize the transverse shear from the in-plan stresses. Thus from:

(V.11)
$$\begin{cases} \partial_\beta \sigma_{\alpha\beta} + \partial_3 \sigma_{\alpha 3} = 0 \ \ \text{on} \ \ \Omega^\varepsilon \\ \sigma_{\alpha 3} = 0 \ \ \text{on} \ \ \Gamma^\varepsilon_\pm \end{cases}$$

we forget any kind of loading other than transverse pressure $g_3^\pm$ and in-plan lateral forces applied on the boundary Γ_1^ε (see Figure V.1). This enables one to derive the following expression:

$$\sigma_{\alpha 3} = - \int_{-\varepsilon}^{x_3} \partial_\beta \sigma_{\alpha\beta} \ \text{on} \ \omega_d \ \text{and} \ \omega_s .$$

This formula is valid even for $x_3 > z$ or $x_3 < z$ because the transverse shear is continuous at the interface between the two portions of the delaminated area.

Numerically speaking, the computation of $\sigma_{\alpha 3}$ is not easy because it involves stress derivatives. For instance, if we use piecewise constant approximations for the stresses $\sigma_{\alpha\beta}$, the previous expression for $\sigma_{\alpha 3}$ only involves Dirac distributions at the interface between the elements of a mesh. Therefore a mixed formulation where the transverse shear is an independent variable is suitable.

Remark V.2
The suggested plate model is just an assembly of three sub-plates connected together through relations (V.6). But the non-symmetrical distribution of the composite layers induces a coupling between membrane and bending effects into the two plates above and under the delaminated area. In addition this coupling is increased due to the continuity relations (V.6). ■

Remark V.3
From Stokes formula applied to the variational equations (V.10), and because of the continuity relationships (V.6), one can easily derive the stress continuities at the crack tip, but not continuity for all the components of the stress vector, as we shall see. These relations are (see Figure V.1):

i) continuity of v_α :

$$\left(n_{\alpha\beta} - n^+_{\alpha\beta} - n^-_{\alpha\beta}\right) b_\beta = 0 \text{ along } \gamma_f \text{ for } \alpha = 1, 2$$

ii) continuity of the rotation along γ_f :

$$\left(m_{\alpha\beta} - m^+_{\alpha\beta} - m^-_{\alpha\beta}\right) b_\alpha b_\beta - \left(n^+_{\alpha\beta}\xi^+ + n^-_{\alpha\beta}\xi^-\right) b_\alpha b_\beta = 0$$

iii) continuity of v_3 :

$$\partial_s\left[\left(m_{\alpha\beta} - m^+_{\alpha\beta} - m^-_{\alpha\beta}\right) b_\alpha a_\beta\right] + \partial_\alpha\left(m_{\alpha\beta} - m^+_{\alpha\beta} - m^-_{\alpha\beta}\right) b_\beta$$
$$-\partial_s\left(n^+_{\alpha\beta} b_\beta a_\alpha \xi^+ + n^-_{\alpha\beta} b_\alpha a_\beta \xi^-\right) = 0$$

where s is the curvilinear abscissa along the boundary γ_f. It is worth here again to point out that the three above relations do not imply the continuity of the normal stress at the interface between the three sub-plates (except in average through the thickness).   ■

V.4   The three-dimensional energy release rate

Let us first recall the mechanical framework which leads to the definition of the energy release rate as a meaningful quantity in fracture mechanics. The elastic energy of a structure at the equilibrium is a function of the various parameters which define the structure. For instance, and basically for our case, the crack tip position is such a parameter. Let us denote by $u = (u_i)$ the three-dimensional displacements in the structure and by $\sigma = (\sigma_{ij})$ the stress field. Then the elastic energy is:

(V.12)
$$J = \frac{1}{2}\int_\Omega \sigma_{ij}\gamma_{ij}(u) - I(u)$$

$\gamma_{ij}(u) = \frac{1}{2}\left(\partial_i u_j + \partial_j u_i\right)$ being the strain field and $I(u)$ the external energy due to the applied forces. The basic idea due to A.A. Griffith [20] consists in assuming that the amount of energy that the structure can spend in the crack evolution is the corresponding variation of the elastic energy the expression of which is given in (V.12). Thus the derivative (up to the sign) of the elastic energy appears as a thermodynamic force which drives the crack tip evolution. The thermodynamic approach of this statement has been introduced and widely explored by N-

Guyen Q.S. [21]. From a purely technical point of view, the computation of the derivative was performed by Ph. Destuynder and M. Djaoua [22], using the domain derivative methodology as it has been introduced by F. Murat and J. Simon [23]. We recall in the next section the expression of this derivative. See also for the experimental aspects the paper by O'Brien [19] and for another approach the one by Anquez and Nataf [20]

V.4.1   *The energy release rate*

Let us denote by $\theta = (\theta_i)$ a vector field defined on the open set Ω^ε (occupied by the plate) and satisfying the following properties:

i) the restriction of θ to the crack tip curve represents a virtual displacement of it;

ii) the support of θ is a neighbourhood of the crack tip curve;

iii) the vector field θ is everywhere parallel to the delaminated area which is also parallel to medium surface ω of the plate;

iv) the components θ_α of θ are independent of the coordinate x_3 (the one which describes the thickness of the plate).

To any value, say η, of a small parameter, we associate the new open set Ω^ε_η which is like Ω^ε except that the position of the crack tip curve has moved in the following manner:

For any $M \in \Omega^\varepsilon$ we set:

$$F^\eta(M) = M + \eta\,\theta(M) \in \Omega^\varepsilon_\eta, \; \eta \in \mathbb{R}^+ .$$

Then we associate to the open set Ω^ε_η the elastic model the solution of which is denoted by (σ^η, u^η). Obviously, for $\eta = 0$, the solution (σ^η, u^η) corresponds to the one of the model set over the open set Ω^ε. Then we introduce the elastic energy of the plate where the delaminated area has moved in the displacement field $\eta\,\theta$:

$$J^\eta = \frac{1}{2} \int_{\Omega^\varepsilon_\eta} \sigma^\eta{}_{ij} \partial_i u^\eta{}_j - l(u)$$

where coordinate derivatives: $\partial_i(\bullet)$ has to be understood as derivative with respect to the coordinates:

$$x^\eta_i = x_i + \eta\,\theta_i$$

The derivative of the elastic energy J^η with respect to the crack tip position – say Γ_f – in the displacement field θ is defined by:

$$G = - \lim_{\eta \to 0} \frac{J^\eta - J^0}{\eta} \quad .$$

The calculation of G can be performed analytically. If one notices that the partial derivative of J^η with respect to u^η is zero (u^η minimizes the elastic energy), it is sufficient to consider the explicit dependence on the parameter η. This leads to the following expression which was introduced by Ph. Destuynder and M. Djaoua [22]:

(V.13) $\qquad G = -\dfrac{1}{2} \displaystyle\int_{\Omega^\varepsilon} \sigma_{ij} \partial_i u_j \partial_\alpha \theta_\alpha \; + \int_{\Omega^\varepsilon} \sigma_{ik} \partial_\alpha u_k \partial_i \theta_\alpha$

(the summation is assumed over repeated indices from one to three concerning Latin indices and from one to two for Greek indices). As a matter of fact, the integrals over Ω^ε which appear in the expression of G are limited to a neighbourhood of the crack tip Γ_f because the support of θ is restricted to such a neighbourhood. With another respect, using Stokes formula, one can prove that G only depends on the value of the vector field θ on the crack tip Γ_f, and more precisely on the normal component of θ along Γ_f. But the formula (V.13) is more useful for our purpose. First of all let us notice that G can also be written:

(V.14) $\; G = -\dfrac{1}{2} \displaystyle\int_{\Omega^\varepsilon} \sigma_{ij} \partial_i u_j \partial_\alpha \theta_\alpha \; + \int_{\Omega^\varepsilon} \sigma_{\mu\beta} \partial_\alpha u_\beta \partial_\mu \theta_\alpha \; + \int_{\Omega^\varepsilon} \sigma_{\mu 3} \partial_\alpha u_3 \partial_\mu \theta_\alpha$

In order to derive an approximation for G from the plate model solution described in section V.3, it could be natural to introduce directly in (V.14) the corresponding approximation of (σ, u). Unfortunately, this is not the right way, and this would lead to a false expression when the thickness of the plate is very small as compared to the other dimensions in the structure. The limit value of the energy release rate is different from the previous quantity. Let us explain this phenomenon from a mechanical point of view. The transverse shear stress $\sigma_{\alpha 3}$ ($\alpha = 1, 2$) is defined in the Kirchhoff-Love model as a distribution. Thus it can be represented as a linear form on the internal displacement fields, but also with a concentrated force along the free edges of the plate. As a matter of fact, this last term is zero for the three-dimensional model (boundary condition on a free edge) but it is not zero for the plate model (due to the strange boundary condition on a free edge). The next section gives a justification of the right energy release rate for plates based on mechanical considerations and with reference to a mathematical proof which was detailed in Ph. Destuynder - Th. Nevers [1].

V.4.2   The energy release rate for delaminated plates

The justification of the three-dimensional model for plates introduced in section V.3 could be completely justified in a mathematical framework using the so-called asymptotic method. But this would complicate our presentation. In order to be simple we try to restrict our developments to the minimum of mathematics, refering the reader to [1] for a complete analysis.

Let us first recall the basic assumptions which have been used in section V.3:

i) The transverse shear energy can be neglected compared to the membrane or bending one;
ii) The plane stresses are well approximated (in energy norm) by the solution to the Kirchhoff-Love model;
iii) Displacement and rotation fields are well approximated by the Kirchhoff-Love model.

From the three above assumptions, we suggest an approximation of the energy release rate which could be justified in a mathematical framework using the asymptotic method. The details of such a proof are explicited in Ph. Destuynder and Th. Nevers [1]; but the following considerations also make it easy to understand it works.

First of all, the Principle of Virtual Work enables one to write that:

$$(V.15) \qquad \forall \ v = (v_\alpha), \quad \int_{\Omega^\varepsilon} \sigma_{\alpha\beta} \partial_\alpha v_\beta + \int_{\Omega^\varepsilon} \sigma_{\alpha 3} \partial_3 v_\alpha \ = \ l(v)$$

where v is a virtual displacement field which satisfies the kinematical boundary condition and such that $v_\alpha \in H^1(\Omega^\varepsilon)$. Let us point out that for sake of simplicity we do not make explicit the loading term $l(v)$.

Then, from the constitutive relationship between shear stress and shear strain, one has:

$$4 S_{\alpha 3 \beta 3} \sigma_{\beta 3} = \partial_\alpha u_3 + \partial_3 u_\alpha \quad \text{for } \alpha = 1, 2,$$

$S_{\alpha 3 \beta 3}$ being the components of the compliance tensor. Thus by eliminating $\partial_\alpha u_3$, we obtain the following equality (for a term which appears in the expression of G):

$$\int_{\Omega^\varepsilon} \sigma_{\mu 3} \partial_\alpha u_3 \partial_\mu \theta_\alpha \ = \ \int_{\Omega^\varepsilon} 4 S_{\alpha 3 \beta 3} \sigma_{\mu 3} \sigma_{\alpha 3} \ \partial_\mu \theta_\beta \ - \ \int_{\Omega^\varepsilon} \sigma_{\mu 3} \partial_3 (u_\alpha \partial_\mu \theta_\alpha)$$

and from the Principle of Virtual Work recalled in (V.15), we deduce that:

$$\int_{\Omega^{\varepsilon}} \sigma_{\mu 3} \partial_{\alpha} u_3 \partial_{\mu} \theta_{\alpha} = \int_{\Omega^{\varepsilon}} 4 S_{\alpha 3 \beta 3} \sigma_{\mu 3} \sigma_{\alpha 3} \partial_{\mu} \theta_{\beta} + \int_{\Omega^{\varepsilon}} \sigma_{\alpha\beta} \partial_{\alpha} (u_{\lambda} \partial_{\beta} \theta_{\lambda}).$$

The loading term has disappeared from the above formula because we assume that no forces were applied in the vicinity of the crack tip which precisely corresponds to the support of the vector field θ. This is not a restriction but only a simplification in the calculations. Let us finally consider the new expression of the energy release rate that we have derived, for the three-dimensional case :

(V.16)
$$G = -\frac{1}{2} \int_{\Omega^{\varepsilon}} \sigma_{ij} \partial_i u_j \partial_{\alpha} \theta_{\alpha} + 4 \int_{\Omega^{\varepsilon}} S_{\alpha 3 \beta 3} \sigma_{\mu 3} \sigma_{\alpha 3} \partial_{\mu} \theta_{\beta}$$
$$+ \int_{\Omega^{\varepsilon}} \sigma_{\alpha\beta} \partial_{\lambda} u_{\beta} \partial_{\alpha} \theta_{\lambda} + \int_{\Omega^{\varepsilon}} \sigma_{\alpha\beta} \partial_{\beta} (u_{\mu} \partial_{\alpha} \theta_{\mu}).$$

Obviously, (V.16) is equivalent to (V.14) as far as the three-dimensional solution (σ , u) is plugged into the expression of G. But the approximation of G using the three assumptions i), ii) and iii) mentioned earlier can only be applied to (V.16) and not to (V.14). The big difficulty in (V.14) is the presence of transverse shear stress in terms which cannot be cancelled (and should not !). Thus, from (V.16), we can eliminate the second term, which represents a transverse shear energy, compared to the first one, which represents a membrane and bending energy. In the three remaining terms the approximation of $\sigma_{\alpha\beta}$ and u_i from the plate model can be plugged into (V.16) and we obtain:

(V.17)
$$G^{\circ} = -\frac{1}{2} \int_{\Omega^{\varepsilon}} \sigma_{\lambda\beta} \partial_{\lambda} u_{\beta} \partial_{\alpha} \theta_{\alpha} + \int_{\Omega^{\varepsilon}} \sigma_{\alpha\beta} \partial_{\beta} (u_{\mu} \partial_{\alpha} \theta_{\mu})$$
$$+ \int_{\Omega^{\varepsilon}} \sigma_{\alpha\beta} \partial_{\lambda} u_{\beta} \partial_{\alpha} \theta_{\lambda}$$

where $(\sigma_{\alpha\beta}, u_{\alpha})$ is the plate model solution as it has been defined in section V.3.

As a matter of fact, since the plate model contains three sub-plates, it is interesting to split G° into the three corresponding contributions (the safe plate and the upper and lower parts with respect to the delaminated area). Thus we have, with self-explanatory notations:

$$G^{\circ} = G_s + G_+ + G_-.$$

with the following expressions (the notations introduced in section V.3 are used):

(V.18)

$$G_s = -\frac{1}{2} \int_{\omega_s} n_{\alpha\beta} \partial_\alpha \underline{u}_\beta \partial_\lambda \theta_\lambda + \int_{\omega_s} n_{\alpha\beta} \partial_\lambda \underline{u}_\beta \partial_\alpha \theta_\lambda$$
$$+ \int_{\omega_s} n_{\alpha\beta} \partial_\beta \left(\underline{u}_\mu \partial_\alpha \theta_\mu \right) + \frac{1}{2} \int_{\omega_s} m_{\alpha\beta} \partial_{\alpha\beta} u_3 \partial_\lambda \theta_\lambda$$
$$- \int_{\omega_s} m_{\alpha\beta} \partial_{\alpha\lambda} u_3 \partial_\beta \theta_\lambda - \int_{\omega_s} m_{\alpha\beta} \partial_\beta \left(\partial_\mu u_3 \partial_\alpha \theta_\mu \right)$$

(V.19)

$$G_\pm = -\frac{1}{2} \int_{\omega_d} n_{\alpha\beta}^\pm \partial_\alpha \underline{u}_\beta^\pm \partial_\lambda \theta_\lambda + \int_{\omega_d} n_{\alpha\beta}^\pm \partial_\lambda \underline{u}_\beta^\pm \partial_\alpha \theta_\lambda$$
$$+ \int_{\omega_d} n_{\alpha\beta}^\pm \partial_\beta \left(\underline{u}_\mu^\pm \partial_\alpha \theta_\lambda \right) + \frac{1}{2} \int_{\omega_d} m_{\alpha\beta}^\pm \partial_{\alpha\beta} u_3^\pm \partial_\lambda \theta_\lambda$$
$$- \int_{\omega_d} m_{\alpha\beta}^\pm \partial_{\lambda\beta} u_3^\pm \partial_\alpha \theta_\lambda - \int_{\omega_d} m_{\alpha\beta}^\pm \partial_\beta \left(\partial_\mu u_3^\pm \partial_\alpha \theta_\mu \right)$$

where ω_s (respectively ω_d) denotes the safe part (respectively delaminated part) of the plate. Hence there are eighteen terms to be computed from the Kirchhoff-Love plate model and each one is only dependent of quantities which are perfectly well-defined in the plate model which has been described in section V.3. It is very important to notice that the obtained formula is not obvious and could not be derived from the only plate elastic energy. The difficulty would appear in the term involving the transverse shear stress. As a matter of fact, one could suggest at least two other expressions for the energy release rate.

• The first one is obtained by taking the derivative of the elastic energy of the plate model with respect to the delamination curve. In such a method, the term

$$\int_{\Omega^\varepsilon} \sigma_{\alpha\beta} \partial_\beta \left(u_\mu \partial_\alpha \theta_\mu \right)$$

would be eliminated. As a matter of fact, it has the physical meaning of the transverse shear stress work in the crack evolution.

• The second formulation which could be suggested consists in approximating the expression (V.14) directly by introducing the plate model solution. For instance, the transverse shear stress would be estimated from the Kirchhoff-Love model by integrating the three-dimensional equilibrium equations (see Chapter I). Then the energy release rate is approximated by the following expression:

$$G^\circ = -\frac{1}{2} \int_{\Omega^\varepsilon} \sigma_{\lambda\beta} \partial_\alpha u_\beta \partial_\lambda \theta_\lambda + \int_{\Omega^\varepsilon} \sigma_{\mu\beta} \partial_\alpha u_\beta \partial_\mu \theta_\alpha + \int_{\Omega^\varepsilon} \sigma_{\mu3} \partial_\alpha u_3 \partial_\mu \theta_\alpha \; .$$

By expliciting the various terms, we obtain:

(V.20)
$$G^\circ = G_p + G_+ + G_-$$

with:

(V.21)
$$G_p = -\frac{1}{2} \int_{\omega_s} n_{\alpha\beta} \partial_\alpha \underline{u}_\beta \partial_\lambda \theta_\lambda + \int_{\omega_s} n_{\alpha\beta} \partial_\lambda \underline{u}_\beta \partial_\alpha \theta_\lambda$$
$$+ \frac{1}{2} \int_{\omega_s} m_{\alpha\beta} \partial_{\alpha\beta} u_3 \partial_\lambda \theta_\lambda - \int_{\omega_s} m_{\alpha\beta} \partial_{\alpha\beta} u_3 \partial_\alpha \theta_\lambda + \int_{\omega_s} Q_\mu \partial_\alpha u_3 \partial_\mu \theta_\alpha$$

(V.22)
$$G^\pm = -\frac{1}{2} \int_{\omega_d} n^\pm_{\alpha\beta} \partial_\alpha \underline{u}^\pm_\beta \partial_\lambda \theta_\lambda + \int_{\omega_d} n^\pm_{\alpha\beta} \partial_\lambda \underline{u}^\pm_\beta \partial_\alpha \theta_\lambda$$
$$+ \frac{1}{2} \int_{\omega_d} m^\pm_{\alpha\beta} \partial_{\alpha\beta} u^\pm_3 \partial_\lambda \theta_\lambda - \int_{\omega_d} m^\pm_{\alpha\beta} \partial_{\alpha\lambda} u^\pm_3 \partial_\beta \theta_\lambda + \int_{\omega_d} Q^\pm_\mu \partial_\alpha u^\pm_3 \partial_\mu \theta_\alpha$$

with the notation:

on ω_s :
$$Q_\alpha = \int_{-\varepsilon}^{\varepsilon} \sigma_{\alpha 3} \, dx_3$$

on ω_d :
$$\left|
\begin{aligned}
Q^-_\alpha &= \int_{-\varepsilon}^{z} \sigma_{\alpha 3} \, dx_3 \\
Q^+_\alpha &= \int_{z}^{\varepsilon} \sigma_{\alpha 3} \, dx_3
\end{aligned}
\right.$$

The main difference between (V.18) - (V.19) on the one hand and (V.21) - (V.22) on the other hand lies on the contribution of the transverse shear stress. As a matter of fact, it is possible to use an integration by part in order to compare these terms. Because one has for instance on the safe part – say ω_s – of the medium surface of the plate (see Chapter I):

$$Q_\alpha = \int_{-\varepsilon}^{\varepsilon} \sigma_{\alpha 3} = \int_{-\varepsilon}^{\varepsilon} \int_{-\varepsilon}^{x_3} \partial_\beta \sigma_{\alpha \beta} \ ,$$

assuming for sake of simplicity that $g_\alpha^{\cdot} = 0$, but from (see Chapter I):

$$\sigma^{o}_{\alpha \beta} = \frac{\left[n_{\alpha \beta} + x_3 \, m_{\alpha \beta} \right]}{2 \, \varepsilon}$$

we deduce that:

$$Q_\alpha = \frac{\varepsilon^2}{3} \partial_\beta m_{\alpha \beta} - \varepsilon \partial_\beta n_{\alpha \beta}$$

Hence:

$$\int_{\omega_s} Q_\mu \partial_\alpha u_3 \partial_\mu \theta_\alpha = \int_{\omega_s} \left[\frac{\varepsilon^2}{3} \partial_\beta m_{\mu \beta} - \varepsilon \partial_\beta n_{\mu \beta} \right] \partial_\alpha u_3 \partial_\mu \theta_\alpha$$

or else:

$$\int_{\omega_s} Q_\mu \partial_\alpha u_3 \partial_\mu \theta_\alpha = \int_{\partial \omega_s} \left[\frac{\varepsilon^2}{3} m_{\mu \beta} - \varepsilon n_{\mu \beta} \right] b_\beta \partial_\alpha u_3 \partial_\mu \theta_\alpha$$

$$- \int_{\omega_s} \left[\frac{\varepsilon^2}{3} m_{\mu \beta} - \varepsilon n_{\mu \beta} \right] \partial_\beta \left(\partial_\alpha u_3 \partial_\mu \theta_\alpha \right)$$

Thus if we notice that in the plate model one has:

$$u_\alpha = - x_3 \partial_\alpha u_3 \, ,$$

we can write:

$$\int_{\omega_s} Q_\mu \partial_\alpha u_3 \partial_\mu \theta_\alpha = \int_{\partial \omega_s} \int_{-\varepsilon}^{\varepsilon} \sigma_{\mu \beta} b_\beta u_\alpha \partial_\mu \theta_\alpha + \int_{-\varepsilon}^{\varepsilon} \sigma_{\mu \beta} \partial_\beta \left(u_\alpha \partial_\mu \theta_\alpha \right) \ .$$

Hence the difference between (V.18)-(V.19) and (V.21)-(V.22) lies on a boundary term. Because the support of θ_α is restricted to a vicinity of the delamination tip, this boundary term is also restricted to the crack tip. More precisely in a Kirchhoff-Love model one has not the continuity of the normal stress $\sigma_{\mu \beta} b_\beta$ (see Remark V.3). Hence the boundary term is not zero and the two expressions for the energy release rate are not identical.

Let us now underline that the right expression is the one given at (V.18)-(V.19). This can be proved in a mathematical framework [1]. But one can also justify this result from general considerations. The basic point is that the transverse shear stress obtained from the Kirchhoff-Love model is a distribution lying in the space $H^{-1}(\omega)$ and therefore it cannot be used in an integral over ω. The next point is that the term

$$\int_{\Omega^\varepsilon} \sigma_{\alpha\beta} \partial_\beta \left(u_\lambda \partial_\alpha \theta_\lambda \right)$$

is asymptotically stable when the thickness 2ε of the plate tends to zero (see the reference [1]). Finally these two remarks prop the validity of (V.18)-(V.19).

The numerical solution of the plate model is very difficult as far as one looks for a good accuracy on the transverse shear terms. As a matter of fact the two numerical formulations of Chapter III have been used and compared. As a matter of fact we choose a simple expression of the vector field θ which is piecewise linear and based on the finite element mesh. But in order to permit the computation of G^o for the Kirchhoff-Love model, we used an analytical definition of the components θ_α (piecewise cubic and based on the finite element mesh), so that the second order derivatives are smooth enough even if they are not exactly continuous at the boundary between elements.

V.5   The mechanical example and the numerical method

The study which is presented in this section has been carried out by Thierry Nevers [2] in a first step and then prolongated by Mekki Ayadi [24] in his doctoral dissertation in 1990. The example was suggested by the "Laboratoire Central de l'Aérospatiale" . An experiment has been performed by this company, which proved that the numerical values obtained by Th. Nevers and M. Ayadi are very correct.

V.5.1   The specimen studied

The plate that we consider here is manufactured in an homogeneous and isotropical material. It has a rectangular shape, as shown on Figure V.4.

Figure V.4 - The specimen used for the computations of G (carbon-epoxy)

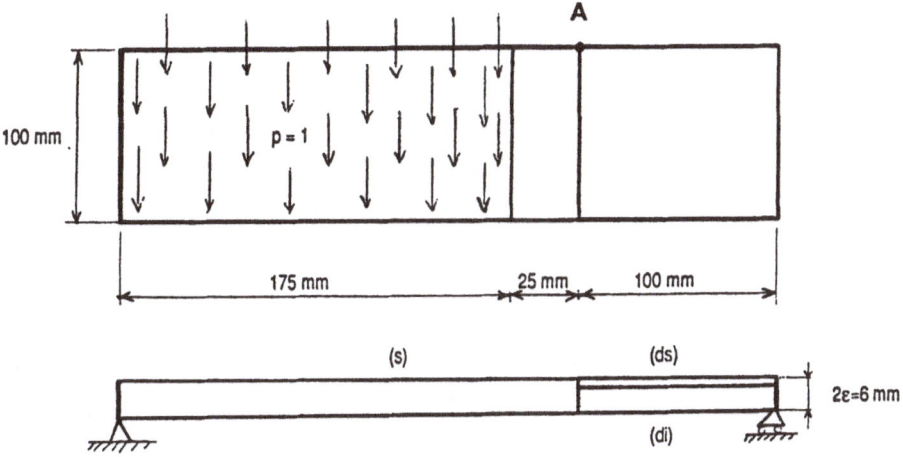

Figure V.5 - The mesh used in the finite element computations

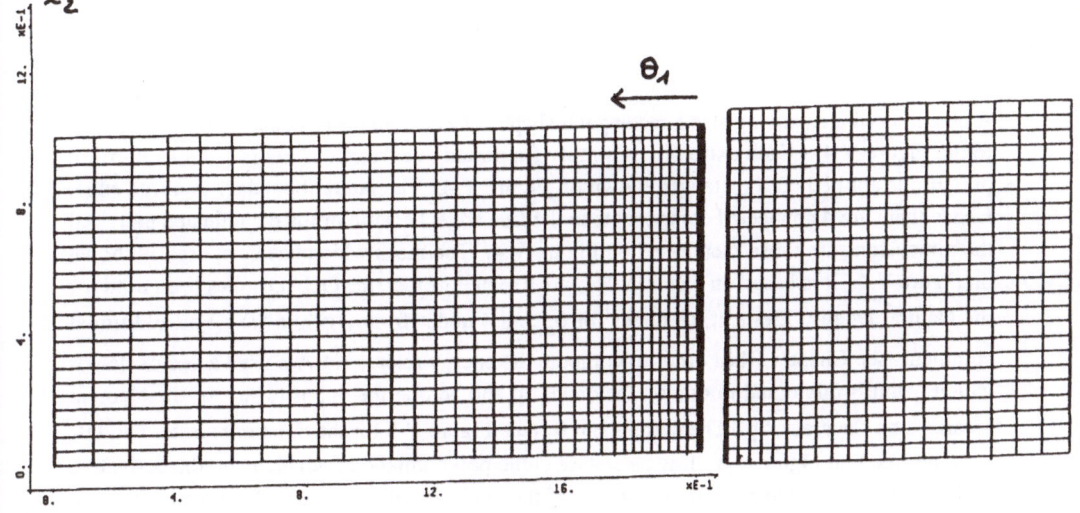

Figure V.6 - The computation of G versus the crack length

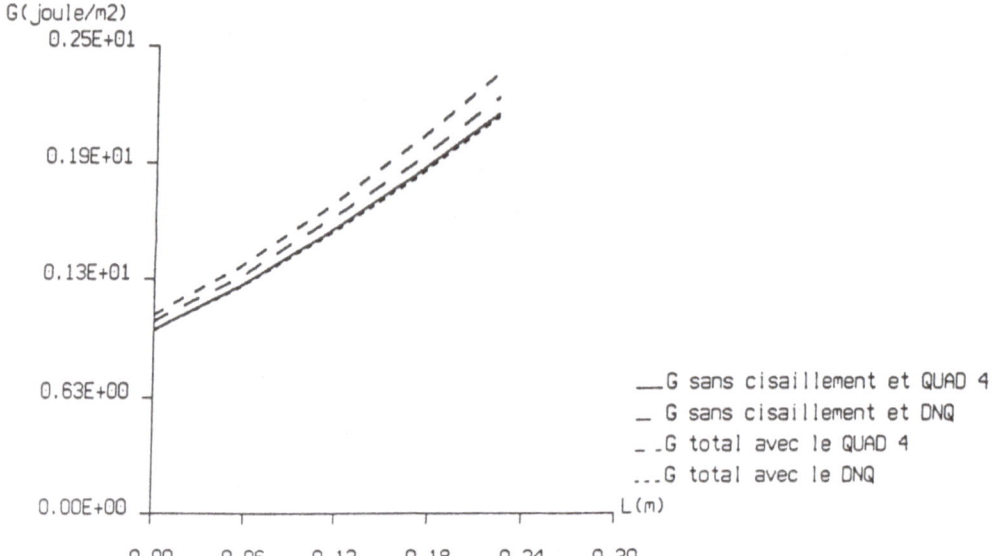

A delaminated area having also a rectangular shape is located on the right part of it (with respect to the Figure V.4). Each extremity is simply supported , i.e. rotations are free, and a uniform pressure is applied on a rectangle shown on Figure V.4. The lateral boundary is a free edge. Hence there are 4 connected components to the free edge, because of the delaminated part. All the dimensions are those mentioned on Figure V.4. The mesh which has been used is shown on Figure V.4. It is based on quadrangles. As a matter of fact, the QUAD 4 has been used for the Natural Duality Element (8 nodes for θ_α and 4 nodes for u_3, ϕ and ψ). For additional comments see Chapter III.In order to compute the energy release rate we used a vector field θ which is uniformly equal to one on the crack tip and zero at the opposite side of the element localized on the crack tip. The component θ_2 (see Figure V.5 for the notations) is zero and θ_1 is represented on Figure V.5. It is piecewise cubic polynomial on each element, the derivative at the nodes being zero at both extremities of the element with respect to the direction x_1 (see Figure V.5). Then the values of the energy release rate is represented on Figure V.6 (versus the crack length). Let us comment the results. First of all the QUAD4 and the DNQ elements lead to quite similar curves. But it seems that the contribution of the transverse shear stress is quite small. As it is very badly approximated in the QUAD4 formulation, it is better to forget it from the expression of the energy release rate. Obviously the example tested here is not sufficient. But it has the advantage of being very simple. Another example has been tested and compared to the experiment. It has been extensively treated by Th. Nevers in his thesis [25]. In this case the finite element mesh used is formed of triangles and we consider a four-layer composite plate. Each layer is made of glass fibers glued with epoxy; the plate is symmetrical with respect to its medium surface. The orientation of the fibers is shown on Figure V.7.

Figure V.7 The plate used for a comparison between modelling and experiment (glass-epoxy)

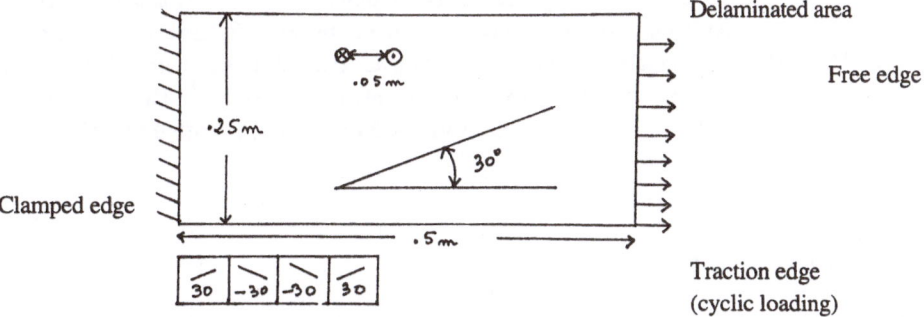

Orientation of the fibers through the thickness (four layers): 30/-30/-30/30

The left boundary is clamped (see also Figure V.8). A traction is applied on the right (see Figures V.7 and V.8). A delaminated area has been developed by a fatigue loading (the frequency is 1 Hz). The initial damaged area looks like a triangle. As a matter of fact the layers of glass fibers were built by sticking together several strips. The junction between two strips has then been broken along the line A-B on Figure V.7. Therefore the crack tip is localized on B-C (the free edge B-C-A being also broken). The characteristics of the used material are the following:

$$\begin{cases} E_L \;=\; 70 \text{ GPa} \\ E_T \;=\; 9 \text{ GPa} \\ G_{LT} =\; 4 \text{ GPa} \end{cases}$$ (Giga Pascals)   (longitudinal modulus)
"   (transverse modulus)
"   (shear modulus)

The stiffness tensors are then computed from these physical values. The thickness of the plate and its dimensions are those indicated on Figure V.7. Then the energy release rate is computed at each vertex of the crack tip. Practically, we used a θ vector field (see formulae (V.21) and (V.22)) which is equal to the shape functions (in the P1-Lagrange approximation) of the node. Then G is normalized by the unit area creation:

$$ - \int_\omega \theta_\alpha \, b_\alpha = 1 $$

where b=(b$_\alpha$) denotes the unit normal to the crack tip. Thus G is computed as a vectorfield along the crack tip. The obtained values are drawn on Figure V.10 while the mesh used is the one of Figure V.8 for the "safe" part" of the plate and Figure V.9 for the delaminated area. Then the crack tip has been moved proportionally to the local value of G, the maximum length being the element size. A remeshing was then used (see FigureV.11) and a new computation led to a new value of the energy release rate along the new crack tip. Then the crack tip is moved again proportionally to the magnitude of the local energy release rate (and in the direction of the vector G). This iterative process is reproduced several times. Few steps are shown on Figures V.12, V.13, V.14 and the last one should be compared to the experimental results shown on Figure V.15 The full computation was performed on an **ALLIANT-FX8** with 48MB of main memory. The C.P.U. time was about five hours, mainly due to overlays. On a more recent architecture with a larger memory like the CONVEX-C3 or the CRAY-YMP the C.P.U. time is considerably reduced.

V.6   Concluding remarks

The goal of this chapter was to prove on a realistic example that a very accurate analysis and precise finite element are necessary for modelling a new mechanical problem and they are numerous in plates and shells. But more than a mathematical justification, the physical experiment is the only final judge of a theory. In the last example considered in this chapter, the comparison is quite satisfying. This proves that the concept of energy release rate, as we defined it, is well adapted to the understanding of delamination process. But the numerical implementation (i.e. the programming) has to be performed very carefully in order to fight against the ill conditioning of the model.

Figure V.8 - The initial configuration used in the experimental test

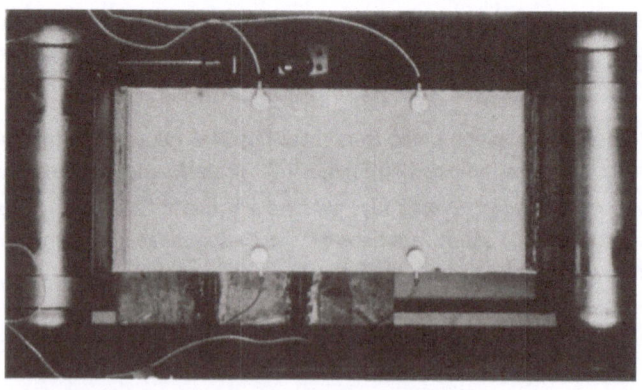

Figure V.9 - The finite element mesh used in the computation (delaminated area)

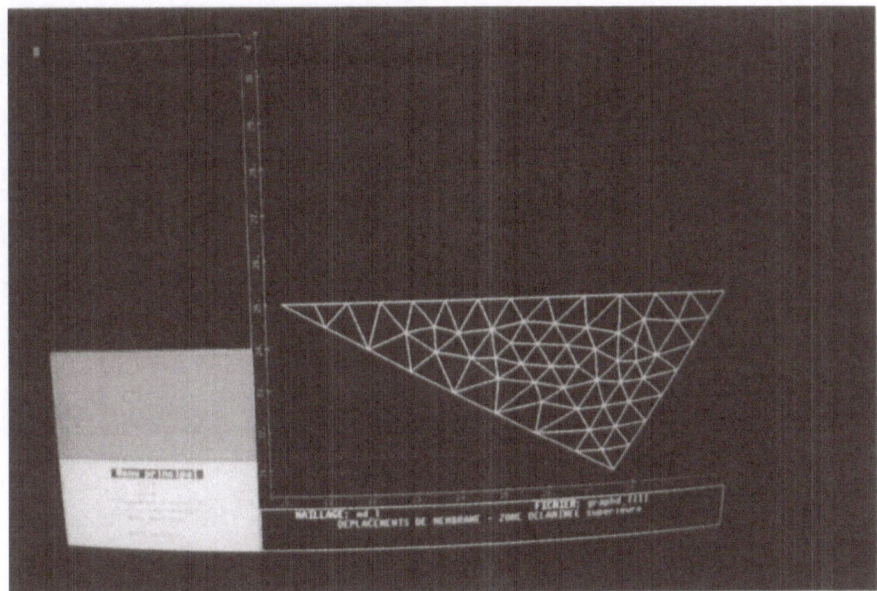

Figure V.10 - The mesh of the safe part of the plate

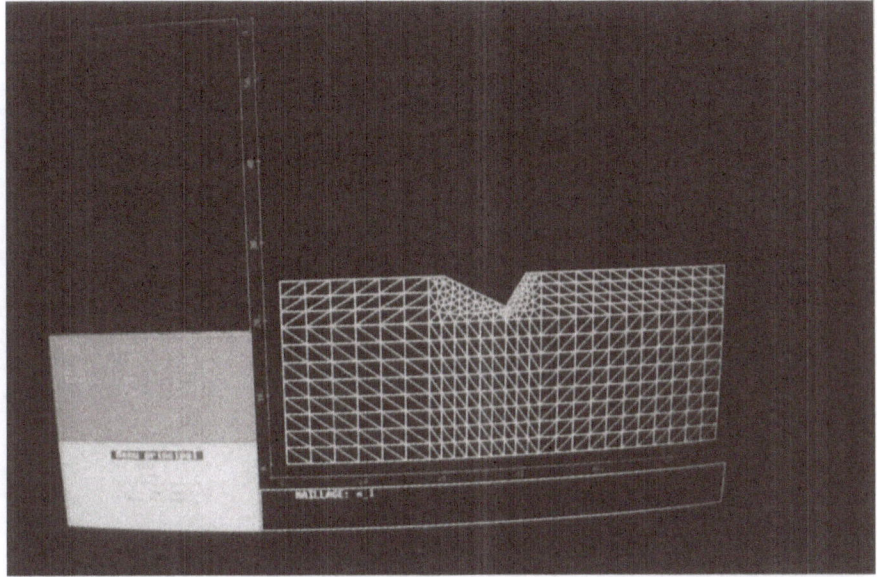

Figure V.11 - New remeshing mesh after five iterations (and the energy release rate)

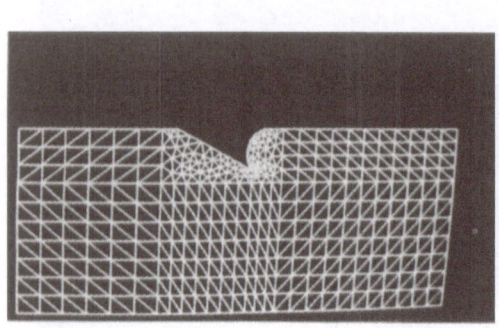
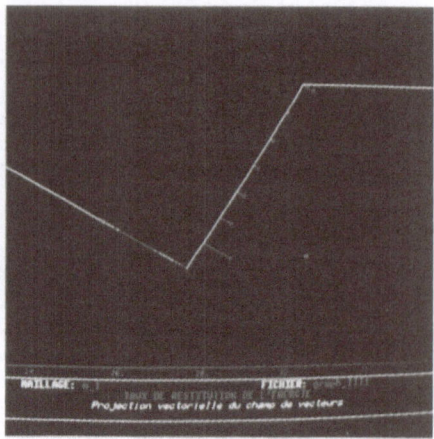

Figure V.12 - New remeshing after ten iterations (and the energy release rate)

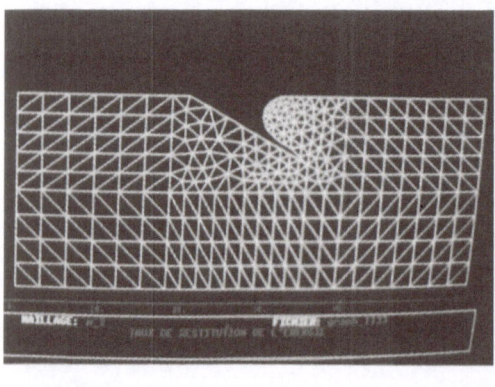
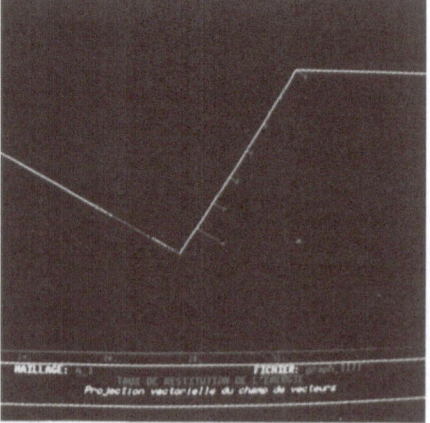

Figure V.13 - Another remeshing after 15 iterations (and the energy release rate)

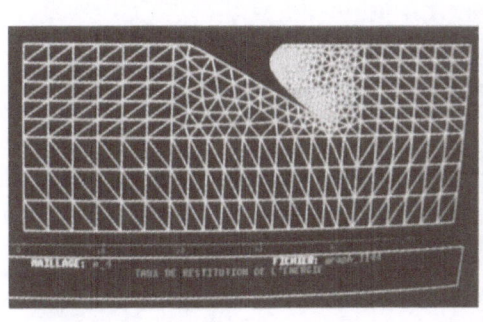
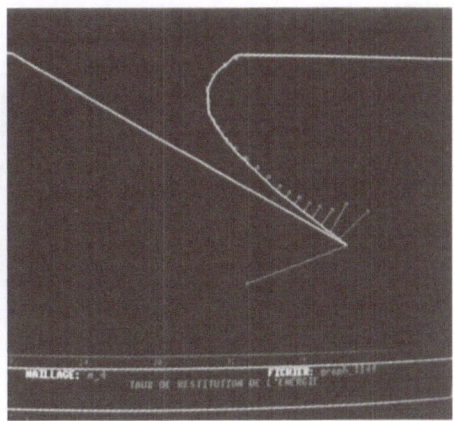

Figure V.14 - New remeshing after 20 iterations (and the energy release rate)

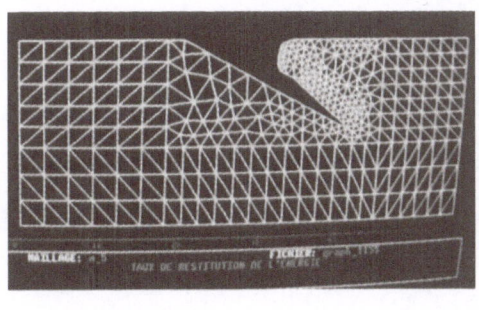
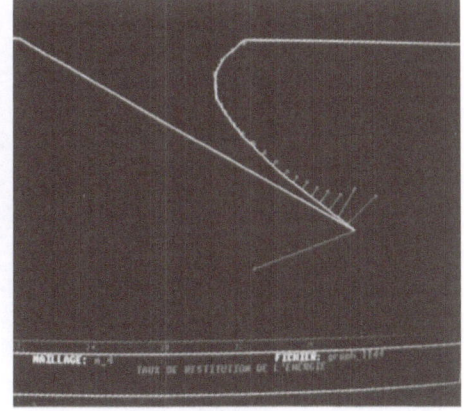

Figure V.15 - The delaminated plate observed experimentally

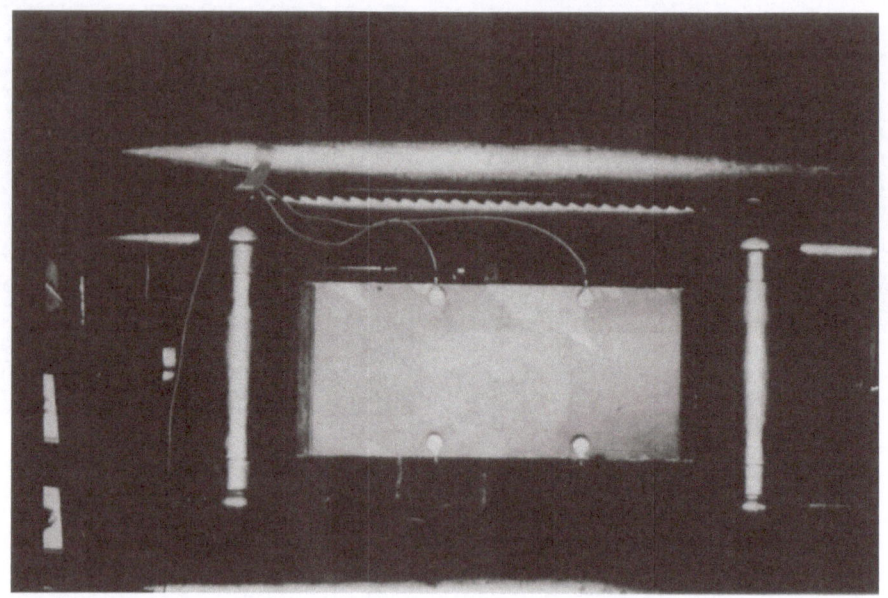

REFERENCES

[1] DESTUYNDER Ph., NEVERS Th., [1987], A model for studyingthe delamination of thin multilayered plates. In Application of multiple scaling mechanics edited by P. G. Ciarlet and E. Sanchez-Palencia, RMA n°4, Masson, Paris.

[2] NEVERS Th., [1986], Modélisation théorique et numérique du délaminage dans les plaques composites, Doctoral Dissertation, Ecole Centrale de Paris.

[3] DESTUYNDER Ph., NEVERS Th., [1987], Un modèle de calcul des forces de délaminage dans les plaques minces multicouches, J.M.T.A., 6 , p. 179-207.

[4] DESTUYNDER Ph., NEVERS Th., [1989], Comparaison des différentes méthodes de calcul des forces de délaminage, Rapport DRET 8734321.

[5] DESTUYNDER Ph., NEVERS Th., [1989], Analysis of damage mechanism using the energy release rate,.Rapport DRET 8734322.

[6] DESTUYNDER Ph., [1982], Sur la propagation des fissures dans les plaques minces en flexion, J.M.T.A., 1, p. 579-594.

[7] GRIFFITH A. A., [1920], Phil. Trans. Soc., London ser.A, p. 221-230.

[8] WANG S.S., [1983], Mechanics for delamination problems in composite materials, vol. 17, p. 210.

[9] WANG S. S., CHOI I., [1982], Boundary layers effects in composite laminates: part 1-2, free edge-sress solution and basic characteristics, J. Appl. Mech., 49, p. 549-560.

[10] WANG A. S. D., CROSSMAN F., [1977], Some new results on the edge effects in symmetric composite laminates, Comput. & Structures, 11, p. 11.

[11] RAJU I., CREWS J., [1981], Interlaminar stress singularities at the free edge in composite laminates, Comput. & Structures, 14, p. 21-28.

[12] ANQUEZ L., BERN A., RENARD J., [1985], Calcul des singularités dans les multicouches, La Recherche Aérospatiale, 1985-2.

[13] DAVET J. L., DESTUYNDER Ph., [1985], Singularités logarithmiques dans les effets de bord d'une plaque en matériaux composites, J.M.T.A., 4, p. 357-373.

[14] DESTUYNDER Ph., OUSSET Y, [1989], Remarks on the damage analysis of adhesive bonds using the energy release rate, Proceedings of the 5th International Symposium on numerical methods in engineering. Computational Mechanics Publications, p. 363-370, Springer-Verlag, Berlin.

[15] DESTUYNDER Ph., OUSSET Y., STACKLER C., [1988], Sur les singularités de contraintes dans les joints collés, J.M.T.A., 7, p. 899.926.

[15] DESTUYNDER Ph., DJAOUA M., LESCURE S., [1983], Quelques remarques sur la mécanique de la rupture élastique, J.M.T.A., 2, p. 113-135.

[16] DESTUYNDER Ph., BONNET E., MICHELIN E., TEMPLIER B., [1988], Computation of stress singularities in edge efects of multilayered composites, La Recherche Aérospatiale, 1988-1.

[17] OUSSET Y., [1990], Modélisation de plaques minces multicouches composite et présentant de délaminages, La Recherche Aérospatiale.

[18] OUSSET Y., ROUDOLPH F., [1992], Un modèle limite de bande mince présentant deux fissures longitudinales, C.R. Acad. Sci. Paris, 314, série II, p. 7-12.

[19] O'BRIEN T. K., [1979], Mixed mode chain-energy release rate effects on edge delamination of composites, NASA Technical Memorendum 84952.

[20] ANQUEZ L., NATAF F., [1987], Délaminage d'un barreau composite soumis à une traction uniforme, J.M.T.A.,6, p. 335-350.

[21] NGUYEN Q. S., [1980], Méthodes énergétiques en mécanique de la rupture, J. Mécanique, vol. 19, n°2.

[22] DESTUYNDER Ph., DJAOUA M., [1981], Sur une interprétation mathématique de l'intégrale de Rice en mécanique de la rupture fragile, Math. Meth. in Appl. Sc., vol.3.

[23] MURAT F, TARTAR L., [1980], Non linear analysis and mechanics, Heriot-Watt symposium, edited by R.J. Knops, Research notes in Mathematics n°27, Pitman Publisher.

[24] AYADI M., [1990], Contribution à l'étude des plaques minces composites délaminées, Doctoral Dissertation, Ecole Centrale de Paris.

A,B

Argyris element p. 114
boundary conditions p. 45, 54, 63, 76, 84
boundary layers p. 120
Brezzi theorem p. 9, 81
bubble function p. 109, 132, 161

C

cache memory p. 198
cantilever plate p. 188
clamped plate p. 189
c^1 elements p. 114
compliance p. 209, 213, 219
composite materials p. 29
conjugate gradient p. 197
connected components p. 76, 82, 91, 95, 99
constitutive relationships p. 6, 63
convergence of the penalty method p. 67
crack propagation p. 212
Crout factorization p. 184, 196, 197
Cuthill-MacKee p. 200

D

deflection p. 83
deformed Lagrange element p. 107, 108
delaminated area p. 213, 217, 221, 226
delamination p.29, 208, 210, 214, 215
deviatoric stress p. 10
domain derivative p. 217

E,F

energy release rate p. 207, 216, 219, 221, 228
error estimates p. 112, 112

experiments p. 224, 227
fibers p. 227
fracture mechanics p. 207
Frenet formula p. 36

G,H,I,J,K,L

Gibbs Poole Stockmeyer p. 199
Hellinger Reissner p. 1, 8
inverse inequality p. 133, 143
Kirchhoff Love p. 1, 9, 10, 12, 14, 15, 223
Korn p. 3
laminateds p. 207
level of a graph p. 199
local equilibrium p. 63

M, N

multilayered plates p. 208
membrane p. 16, 23
mesh p. 105
Mindlin p. 185
mini-element p. 109, 113
Morse storage p. 195
Naghdi p. 1, 20, 22
natural duality p. 73, 87
non conforming element p. 116
non homogeneous boundary conditions p. 86

O, P, Q

opening force p. 212
overstressing p. 29
penalty methods p. 49, 59
plate with a hole p. 197
pointwise load p. 190
Poisson equation p. 43
preconditioning p. 196
quad4 p. 145, 149, 183

R, S

Ramses function p. 109, 132, 161, 164
regular family of triangulations p. 110
Reissner p. 48, 144, 122, 50
remeshing p. 229
simply supported plate p. 1, 191
shear energy p. 149
solution methods p. 144
square plate p. 185
Stokes formula p. 31
stress vector p. 215
surface of a matrix p. 199

T, U, V, W, X, Y, Z

Taylor Hood element p. 183
thermodynamic forces p. 216
trace theorem p. 37
transverse shear p. 18, 122, 134, 215,
218, 220, 221
vector and parallel algorithm p. 195
virtual work p. 4
width of a graph p. 200

Déjà parus dans la même collection

1.

T. Cazenave, A. Haraux
Introduction aux problèmes d'évolution semi-linéaires

2.

P. Joly
Mise en oeuvre de la méthode des éléments finis

3/4.

E. Godlewski, P.-A. Raviart
Hyperbolic systems of conservation laws

5/6.

Ph. Destuynder
Modélisation mécanique des milieux continus

7.

J. C. Nedelec
Notions sur les techniques d'éléments finis

8.

G. Robin
Algorithmique et cryptographie

9.

D. Lamberton, B. Lapeyre
Introduction an calcul stochastique appliqué

10.

C. Bernardi, Y. Maday
Approximations spectrales de problèmes
aux limites elliptiques

11.

V. Genon-Catalot, D. Picard
Eléments de statistique asymptotique

12.

P. Dehornoy
Complexité et décidabilité

13.

O. Kavian
Introduction à la théorie des points critiques

14.

A. Bossavit
Électromagnétisme, en vue de la modélisation

15.

R. Kh. Zeytounian
Modélisation asymptotique en mécanique
des fluides newtoniens

16.

D. Bouche, F. Molinet
Méthodes asymptotiques en électromagnétisme

Déjà parus dans la même collection

17.
G. BARLES
Solutions de viscosité des équations de Hamilton-Jacobi

18.
NGUYEN QUOC SON
Stabilité des structures élastiques

19.
F. ROBERT
Les Systèmes Dynamiques Discrets

20.
O. PAPINI, J. WOLFMANN
Algèbre discrète et codes correcteurs

21.
D. COLLOMBIER
Plans d'expérience factoriels

22.
G. GAGNEUX, M. MADAUNE-TORT
Analyse mathématique de modèles
non linéaires de l'ingénierie pétrolière

23.
M. DUFLO
Algorithmes stochastiques

24.
P. DESTUYNDER, M. SALAUN
Mathematical Analysis
of Thin Plate Models